THE SECRET LIFE O

Julie Hill has 25 years environmental experience, including working with governments, businesses, the environmental think-tank Green Alliance, the Environment Agency and the award-winning Eden Project.

JULIE HILL

The Secret Life
of Stuff

A Manual for a new Material World

VINTAGE BOOKS
London

Published by Vintage 2011

2 4 6 8 10 9 7 5 3 1

First published in Great Britain in 2011 by
Vintage
Random House, 20 Vauxhall Bridge Road,
London SW1V 2SA

www.vintage-books.co.uk

Addresses for companies within The Random House Group
Limited can be found at:
www.randomhouse.co.uk/offices.htm

The Random House Group Limited Reg. No. 954009

A CIP catalogue record for this book
is available from the British Library

ISBN 9780099546580

The Random House Group Limited supports The Forest
Stewardship Council (FSC), the leading international forest
certification organisation. All our titles that are printed on
Greenpeace approved FSC certified paper carry the FSC logo. Our
paper procurement policy can be found at
www.rbooks.co.uk/environment

Typeset by Palimpsest Book Production Limited, Falkirk,
Stirlingshire
Printed and bound in Great Britain by
CPI Cox & Wyman, Reading RG1 8EX

To my husband John, and to my boys, Alex and Rory.
And to my Mum.

Acknowledgements

I could not have written this book without my researchers: Simon Inglethorpe who wrote excellent briefings; Claire Thacker who crunched lots of numbers; and Thomas Turnbull who engaged in tireless tracking down of many sources and references. Heartfelt thanks also to Karen Crane, Patrick Mahon, Mary Orr, Mike Petty and Melissa Tombling for invaluable advice.

Thank you to Hannah Bateson, Dave Matthews and Natsuko Matsuoka for the 'letters from the real world'. They were important in bringing me back to earth.

I also want to thank all those who were kind enough to read sections of the text, and in some cases wade through all of it, providing me with many leads and comments. Staff at the Eden Project, the Environment Agency, Green Alliance and WRAP were particularly helpful and engaged, although all those named below commented in a personal rather than organisational capacity:

Michael F. Ashby, Alison Austin, Mark Barthel, Steve Bass, Alain de Botton, Rosie Boycott, Jeannette Buckle, Germana Canzi, Nick Cartwright, Tracy Carty, Chris Chubb, Rebecca Cocking, Karen Crane, Linda Crichton, Paul Davison, Andy Dawe, Caroline Digby, Liz Dixon-Smith, Chris Dow, Jane Evans, Steve Evans, Jack Frost, Rose George, Ray Georgeson, Martin Gibson, Pam Gilder, Nigel Haigh, Richard Girling, Peter Goult, Ian Hetherington, Allan Hill, Susan Hill, Hannah Hislop, Belinda Howell, Jane

James, Keith James, Andrew Jenkins, Peter Jones, Catherine Juckes, Miranda Kavanagh, Tony Kendle, Alan Knight, Michael Landy, Zoe Laughlin, Neal Lawson, Simon Leaf, Paul Leonard, Craig Liddell, Pete Lunn, Tom Macmillan, Julian Maikelm, Dorothy Maxwell, Daniel Miller, Nick Morley, Matthew Neilson, Julian Parfitt, Sara Parkin, Doug Parr, Sandy Pattisson, Michael Pawlyn, Fred Pearce, Georgina Pearman, Rob Pearson, John Pickup, Peter Reineck, John Sanderson, Peter Seggie, Ben Shaw, Tim Smit, Nina Sweet, Caron Thompson, Charles Thwaites, Michael Warhurst, Peter Whitbread-Abrutat, John Williams, Rebecca Willis, Jim Wiltshire, Stephen Wise and David Workman.

I must add that any mistakes or omissions are entirely my responsibility, especially after I have had such expert help from many quarters. Also that I am writing in a purely personal capacity, so any views expressed here do not necessarily reflect those of the organisations I work with, including the Eden Project, the Environment Agency, and Green Alliance.

A special thank you to my editor at Vintage Books, Rowan Yapp, who was superb at boosting my confidence throughout the process.

Finally, it is customary to thank one's family for putting up with the whole exercise, and I do thank them, they were hugely patient and supportive. I'm also tempted to say that they should thank me, because without this book they would not have been treated to many fascinating dinner-table discussions about the origins of plastic or what to do with sewage sludge. But I suspect that may be over-optimistic.

Contents

Part 1: Future and Past

1. A tapestry   3
2. Evie and Ed throw a party, 2040   8
3. Stuff as waste   13

Part 2: The Stuff of Life

4. From the very beginning   33
5. Deconstruct your home   42
6. Exploring the material world   47

Part 3: Why Stuff Matters

7. Nothing comes free   87
8. Where the limits lie   108

Part 4: Us and Stuff

9. The curse of the giant pink pencil   127
10. The myth of the green consumer   139
11. Letters from the real world   162

Part 5: Designing the Future
12. Where do products come from?   179
13. Everything sustainably sourced   187

14. Everything designed for recovery   200
15. Nutrients cycled   210
16. All energy renewable   224
17. Stemming the flow of stuff   239
18. Care with new promises   246

Part 6: From Here to There

19. Brighton and Beyond   255
20. Humans grow up   266

     Notes   298
     Index   345

Part 1
Future and Past

1

A tapestry

I wanted to write this book to tell a story. As I was writing, it seemed more like a tapestry, with a large number of shining, interwoven threads. We humans can achieve greatness when we put our minds to it. We have surrounded ourselves with things that are ingenious and delightful, augmenting the joy that is all around us in the shape of the natural world. But I'm dismayed at how little we understand the complexity of the material world that sustains us, in all its gloriously varied forms and with all its unexpected qualities. We all love stuff – I am no exception – but we rarely know where it comes from and where it goes, and the shadow it casts. We get the bad side of stuff from some shock-horror headlines, but they are sporadic and unconnected. We rarely appreciate the effort that turns materials into the stuff around us, the effort put in by people from all over the world, using nature's riches. We are blessed in many senses, but not in this essential knowledge. We lack insight and literacy; we are oblivious to the secret life of stuff.

I hope to change that. I started work in the environment movement more than twenty-five years ago, and while some things have got better in that time, a lot haven't, so dismay is sometimes an unavoidable component of my life. Yet I remain an unashamed optimist. I am inclined to blame the lack of progress on insufficient political leadership, but perhaps it is us that are unwilling to be led. If that is the case, the upside is that there is ample room for change. I believe that if the secret life of stuff was less secret, we would all buy into change more fervently than we have to date, and we would be pestering governments and businesses to provide us with the means for change. Exploring stuff has made me feel better equipped to ask for the right things. In fact, not just ask for them, but expect them, and expect them as standard. That way, I don't have to agonise over the choices.

I also believe, somewhat contrary to current fashion, that markets need boundaries, and in the case of environmental care, much firmer boundaries than they've yet experienced. As we say to children, it is for everybody's long-term good, and you'll thank me one day. I don't think a deregulatory approach to the economy can possibly give us what we need for the future, and I hope to demonstrate that in the course of the book.

I want to set out a vision of the future, of how we *might* run our material world. Although future-gazing is rarely completely accurate, I don't think that what I am about to sketch is beyond the bounds of possibility. I am going to talk about how we use materials at the moment, and why that is a problem in terms of our planetary boundaries. I also want to consider the role of us as consumers, and why the idea of the 'green consumer' is not getting us very far – what we actually need is to have some choices taken away from us.

When I started to think about writing this book, I looked around for something to illustrate the way in which few of us question the credentials of the things that cross our paths. My sons were watching a rerun of *Dragons' Den*, the popular TV programme where people who have made a lot of money in business decide whether to invest some of it in start-up companies. They subject would-be entrepreneurs to savage questioning and throwaway putdowns in the name of entertainment, and occasionally, very occasionally, they part with their cash and turn a germ of a good idea into a seedling business. They were considering a pitch for light-up disposable drinks glasses. That seemed a good place to start with the secret life of stuff.

The entrepreneur proudly poured champagne into an elegantly shaped glass containing tiny batteries, which upon contact with fluid of any kind produce a glow in a range of seven neon colours. They emit their ethereal light for some eight hours while the bearer floats around like a human glow-worm (presumably the darker the party, the better the effect) and then, when the guests have gone home and dawn approaches ... you simply throw them away with the uneaten crisps and olive pits. Glass, batteries, chemicals and all.

As the Dragons sipped their champagne they asked the usual questions about patents, price, sales and profits. They then had interestingly different reactions to the idea. One Dragon commented on the wastefulness of such a product, and how disgraceful it seemed that the glasses, and their batteries, would end up in landfill. Another was convinced that there were already similar products on the market. But two Dragons liked the product and, unconcerned by the objections put forward, both offered cash, competing with each other to buy into the idea. The inventor walked away with a deal. His response to the waste objection had been that *eventually* the technology would come along to recycle anything.

My hackles were raised by this result. Why had the deal-makers ignored the concerns about something so patently wasteful and unnecessarily energy consuming? My reaction was instinctive so, in the spirit of truth and justice, I decided to examine my indignation in more detail. Just how bad are disposable, light-up drinks glasses really?

This might be a good point to explain what basis I have for passing judgement on 'stuff' anyway. I have spent most of my career working with the influential charity and think tank Green Alliance, for many years working on waste and product policy. Green Alliance started with a handful of individuals in 1978, and there are still less than twenty, and many people are kind enough to remark that we 'punch above our weight'. I also serve on the board of the Environment Agency, the government regulatory body protecting us from pollution and other forms of environmental damage, and I am a non-executive director for the Eden Project, which has introduced me to some of the most creative and enchanting ways of conveying a positive future. More of Eden in the final chapter.

Back to the glasses. A quick Internet search reveals that there are several suppliers of light-up drinks glasses. They are in fact made of plastic (crystal polystyrene), not glass, and they contain small lithium batteries. They light up through a process called electroluminescence – the glasses are composed of a very thin layer of light-emitting phosphor sandwiched between conductors (formed by the liquid when it is poured into the glass) and activated by an electric field from the battery. One site states that the batteries must not be removed from the container, and once they've expired the whole thing must be 'safely disposed of'. That is likely to mean putting them in a black bag and letting the binmen take them to a landfill, just as the enlightened Dragon suspected, but they do in fact qualify as 'waste electronics' (more specifically 'lighting equipment') under the European Union's Waste Electronic and Electrical Equipment, or WEEE, Directive. That means they should really be taken to a local authority recycling site for specialist processing, in order to recover the metals in the battery, and possibly the plastic if the plant is geared up for it. The light lasts for eight hours and when the glass is empty it goes out, enabling it to be stored, so in theory if your guests are both parsimonious and tired you could get two or three parties out of them. In reality, they are likely to be left sitting around holding the dregs of supermarket Chardonnay, glowing feebly until the early hours when the indignity of the bin bag snuffs out their fairy ambience.

So how much resource use do they represent? Each uses around 38 grams of plastic. It is an oil by-product, and as a proportion of all the plastic on the market, would represent a tiny amount. So if 10,000 people have two parties each year using fifty glasses, and don't send them for recycling, they will be contributing 38 tonnes of these products to landfill. That doesn't seem much compared to the estimated 2.5 million tonnes of the real and completely recyclable glass that ends up in landfill every year, more than half of the glass consumed in the UK.

Perhaps rather than asking how bad, we should ask how good? The glasses are beautiful, in a twentieth-century kind of way, maybe. But are they something we'll be proud of in thirty years' time? Will we watch reruns of *Dragons' Den* marvelling at its primitiveness? Let's be guests at the party of the future.

2

Evie and Ed throw a party, 2040

Evie and Ed live in a smallish village in central England. They are well connected to London, where they used to live – the frequent buses and electric-car pooling scheme mean that they can get to the nearest train station at almost any time of day or night to catch one of six trains an hour – but they also have a lively local social life. Like them, most people work part-time and a lot of it from home, so there is time for social gatherings of all sorts. They like having midsummer parties, watching the sky above their garden grow slowly dark, which also means they can enjoy some cool air – average daytime temperatures in summer are now above 30°C. Not that their house is unbearably hot. It has been designed to maximise 'solar gain' in winter by having a glasshouse front, but by shutting this off from the house in summer and using solar energy to assist the carefully designed natural ventilation, they can enjoy comfortable summer temperatures as well. They make all their own energy, of course, with a variety of devices, and feed some to the backup grid as well.

Tonight they have invited a dozen or so friends round for drinks and a 'sneal' – the modern way of eating. Not a huge 'meal' with several courses and inevitably meat-based, as it used to be in the old days. More a series of snacks – mostly vegetable- and fruit-based morsels served over the course of the evening. These have been prepared with great care and imagination by Ed, using mainly their home-grown produce. Evie prefers the quicker method of tuning in to the recipe channel, following a delivery of specially tailored and measured ingredients to match this week's recommendations – little preparation and absolutely no waste.

Evie and the couple's children, Evo and Eden, are fond of decorating the house for parties. First they pick flowers from the garden. They have a small but very productive 'cutting patch' of native and other flowers, fed by their composted kitchen and toilet waste. There is not enough water to use it to flush toilets any more, but no one minds because thanks to new technology, a type of 'digester', home-waste composting is easy and odour-free, and helps to return nutrients to the soil as well as producing useful gas for cooking. If they have surplus flowers they sell them to the local florist's – flowers are rarely imported from other countries these days, the land is needed to grow food for local consumption. Then they get out candles made from beeswax. Bees are now carefully protected and widely kept, and candles made from paraffin wax, like other oil by-products, have become a rarity since oil was made redundant by widespread solar, wave, plant and wind power. They put the candles in the beautiful engraved glass holders that Evie inherited from her grandmother. They have been in the family for generations, have shown little sign of wear, and if they do get chipped the repair shop in the village can grind the glass so cleverly that the damage barely shows. Evie likes the fact that they look old and the slight imperfections speak of their history. Like most of her friends, she rarely has the desire to buy new things unless they fill a real need. For the decorative and frivolous she relies on making things – she has the time thanks to modern working practices – or on heirlooms like these glasses. The tablecloth is one she had made, using cloth also handed down, a pattern taken from the Internet and stitched by the local 'let us make it' shop where almost anything can be turned into almost anything else by a group of craftspeople.

Evie goes to her room to change. Her daytime clothes are made of a clever, plant-based plastic widely used for packaging, but which for clothing has somehow been made to feel like the old-fashioned cotton that was so energy-, water- and pesticide-intensive. It doesn't last as long as cotton, but for fashionable clothes that is not a problem, and when wearers are bored with the style the fabric can be easily recycled. It is also easily cleaned in the waterless washing machine using little energy. So popular are these fabrics that, like the flower growers, the former cotton growers are now growing food. For evening wear, Evie has some heirloom clothes, again passed down from the previous generation, and adapted and repaired by craftspeople. Evie's mother loved high-quality clothes, and so Evie's collection of silks, cashmere and yes, pure, old-fashioned cotton, is the envy of many of her friends.

It is eight o'clock and the guests are arriving. Evie gets out the glasses. Actually, most drinking vessels are not glass any more. Glass was deemed too energy-intensive to be used routinely and possibly lost to the Universal Recycling System – so they are mainly recyclable plant-based plastic. The only glass glasses are heirloom ones that come out on very special occasions. Ed fills them with the local 'champagne'. They start to glow gently, a clever effect produced by genetically engineering the 'plastic' that the glasses are made of, a tweak that doesn't affect its recyclability, and a genetic change that doesn't find its way into the food part of the crop or the wider environment. Evie finds some music on the multi-functional PC that runs the house's comfort systems as well as keeping them connected to Internet, telephone, television and radio services. Like their other electronic devices, it makes and stores its own energy.

The scene is set. The party talks sporadically about the news. Great that the UK is finally reaping the rewards of selling clean energy technology to the Chinese after two decades of investment. Some of them grumble about the price of mobile phones – more than £1,000 each now that the essential metals are in short supply – but since Evie's last one has been around for nearly ten years and gone through only minor upgrades, it is not such a hardship. When it finally stops working, it will go for 'remanufacturing', which means that most of the original resources will be reused. Most of the friends' conversation revolves around their children, their holiday plans – international train travel has just been made even more attractive by the opening of super-high-speed lines across Europe and beyond. One of the guests spills tomato juice on the sofa, but Evie is not too fussed because the furniture leasing company is due to upgrade it soon, and she fancies a new pattern. They will take care of the stains in the course of refurbishing the fabric for someone else.

At the end of the evening, the guests wander home, their way lit by glow-sticks made from the same material as the glasses, and carrying presents of Ed and Evie's flowers and vegetables. They will reciprocate with things they have made or grown when they next host a party. The couple clear up. Food scraps – not much, the guests were hungry, and anyway it is considered very impolite to waste food – go in the composter. The small amount of food packaging that came with the ingredients delivery goes in the composter (all lightweight packaging is now compostable) and the rest is returnable and reusable so is put out to be collected when the next delivery comes. The glasses and utensils are cleaned in the waterless dishwasher and put away. The recycling collections are only necessary monthly – newspapers have been made largely redundant by the Internet and e-readers, packaging is minimal or reusable, batteries have been made obsolete by self-energy-generating devices, and no one would dream of using something as non-renewable and energy-intensive as metals for everyday uses such as drinks containers. Additives to products that made recycling complex, or any substances that could be toxic to humans or the environment if they escape the Universal Recycling System, are a thing of the past.

Anything that doesn't go in the recycling collection or isn't collected for reuse by the various companies providing services to the house has to be taken to the nearby resource depot. That is not such a hardship, since the once-a-month trip can be combined with visiting the bulk commodity store on the same site to stock up on essentials such as grains and loo roll (not paper from trees now, but a plant waste product like most other things of ephemeral use). The cinema and various eateries are also there, using the heat from the community composting plant – not everyone has space for composting at home. There's also the gym, looking into the huge greenhouse where crops are grown on composted waste using heat from the same process. Ed takes going to the gym very seriously (anyway, it's free) but he doesn't need to, since his diet means he's unlikely to put on weight. As he cycles, Ed considers whether the family should move back to London. Life in the city is so pleasant these days, with the huge traffic-free zones, the high-rise but highly convenient apartments with their food-producing hanging gardens and their communal dining and entertainment complexes. There is easy access to work and good schooling, and none of the pollution that he remembers from his childhood. If anything, it beats life in the country.

Pie in the sky? I hope to convince you that it is all within our reach.

3

Stuff as waste

It's the summer of 2006 and I am visiting my mother. She is eighty-four and housebound, but remains resolutely independent. After a cup of tea and a biscuit, I change into dirty clothes, take a deep breath and go out to her driveway to hose out her wheelie bin.

The bin is alive with maggots. Three months earlier, the council decided to change its collections to 'alternate weekly'. Recyclables (recently allocated their own shiny new blue bin) were to be collected one week, and 'residual' wastes, including food, supposedly collected from the usual green bin during the other week.

This could all have gone relatively smoothly for my mum, had she not been on 'assisted collection'. This means that she is not required to wheel her bin to the end of her drive like her neighbours, because the binmen will do her the honour of coming to get it from outside the door.

When the system changed, the assisted collection suddenly stopped. Someone, somewhere, had failed to update a database, or maybe fill in a form. Despite increasingly frantic phone calls by both myself and Mum to the council, her rubbish went uncollected for nearly eight weeks. In my nightmares I saw the abandoned bin as a cross between the primordial soup and Pandora's box, expecting all manner of new life forms and manifestations of evil to emerge from it. At best a health hazard, at worst something that could earn us an ASBO.

Eventually the message got through, and the bin was emptied. There remained in the bottom some sludge and some dispossessed maggots, but it wasn't all that bad, despite sitting there through one of the hottest spells of weather the south of England had experienced for several years. And fortunately my mum didn't generate too much rubbish, living alone and not consuming (or wasting) all that much. Yes, it smelt when you opened the lid, but wheelie-bin lids are surprisingly effective if closed properly. As I stood in the sunshine, hose in hand, I reflected on the considerable irony of this task in relation to my day job. I had recently been defending local councils' records, saying that it wasn't their fault that people were so wasteful, and that companies designed things that couldn't easily be recycled. In fact, that very morning, I had been down to the BBC in White City to broadcast my vision of moving towards 'A Zero Waste UK'.

It was thus a healthy dose of reality to hose out maggots from a bin. I also had to accept that my views had failed to persuade my own mother, since as far as she was concerned, this was all my fault. What was wrong with putting everything in the same bin and having someone remove it, preferably once a week? Life used to be so much simpler . . .

Yes it did. Did this new council recycling initiative represent progress in any sense? To answer that we have to go back in time.

Waste is nothing new. The waste of previous generations is archaeology's stock-in-trade. As American archaeologist and 'garbologist' William Rathje points out in his wonderful book *Rubbish!*, the elaborate temples and other public buildings that civilisations leave behind represent what they want to say about themselves; it is the other stuff they leave that tells us what they were really like. Rathje and his colleagues sift through the debris of ancient peoples to get a picture of their lives, but they also sift contemporary garbage and excavate modern landfills in an attempt to better understand our own pattern of living. This provides ample material for interesting comparisons. The Mayans, for instance, one of the ancient civilisations of Mexico, deposited such mountains of pottery shards that Rathje observes that they could give modern American consumers a run for their money. Many peoples, including the ancient Greeks, didn't even remove waste from their houses. They simply dumped the detritus in the corners and when it all got too much they covered it up with earth to make a new floor, leaving a treasury of historical evidence literally swept under the carpet. If the waste wasn't left in the house, it could be left in the streets and built over, so that cities and towns everywhere gradually rose above their own middens. British journalist Richard Girling has also written a book on rubbish, which begins with a grim account of conditions in British cities up to the Victorian age, when the appalling squalor caused by a combination of commercial, household and bodily wastes forced the creation of local authorities to take matters in hand. The health consequences were inevitable – cholera was rife, and in the middle of the nineteenth century in England, one in seven children did not survive infancy. As Girling observes, even badgers clean their dens. Humans seem to be particularly adept at fouling their own nests.

A century or so on from the Victorians, who did so much to clean up their act, we haven't finished grappling with the problem. What is 'waste'? Waste is something that someone no longer wants to use, whether or not it might still be useful to someone else. There are two important things we know about waste. One is that there appears to be a strong correlation between 'wealth' as measured by gross domestic product (GDP) and waste. The other is that having generated waste, i.e. having decided that something has reached the end of its useful life, even in the modern world there are not very many options for things to do with it. In a sparsely populated world of hunter-gathering people, it might be possible simply to move on and leave it all behind, but settled civilisations have to live with it. Throw it away? Unfortunately, the laws of physics dictate that there is no such thing as 'away'.

Back to the earth

That's not quite true, since nature has a very good take on 'away' – it is called biodegradability. Dead plants and animals get broken down and recycled back into the system. Gardeners experience this every time a pile of smelly kitchen waste and grass clippings miraculously turns into sweet-smelling compost after a few months and can be dug back into the soil where it provides vital strength and nutrients. A lot of what we produce as waste, both from households and industry, falls into the category of biodegradable – food, garden waste, paper, some textiles. If all these are kept separate from non-biodegradable materials they can be composted by a variety of means, even on a very large scale. 'Green' waste, i.e. garden clippings, old Christmas trees and autumn leaves, is now often collected by councils. The waste can be put into huge piles, or 'windrows', in the open air, and gradually turned and sieved to make usable compost. Because of hygiene concerns, kitchen waste is best treated in enclosed compost plants, on specially aerated floors so that it reaches higher temperatures to kill pathogens. For generations, leftover food was fed to animals, but concerns about 'swill' carrying foot-and-mouth, BSE and other diseases has all but put a stop to the practice in the UK, unless the waste is heat-treated first. An alternative for food and animal wastes of all kinds, particularly the more liquid ones, is treatment in a 'digester', a kind of oversized tin can which stews the waste in the absence of oxygen. This produces both useful gas, and liquid fertiliser which is perfectly good to put on soils. Unfortunately, despite its healthy potential to return to the land in the same cycle as nature intended, most biodegradable waste can't easily go down this route because we ruin it by mixing it with non-degradable wastes.

If the waste is not biodegradable (cans, glass, plastics, nappies, batteries, rubble, etc.), nature cannot deal with it at all. Humans have brought non-degradable wastes into the system in a variety of ways: we pull materials out of the earth's crust, we transform them by applying heat, we combine them and we create wholly novel materials such as plastic polymers. What can be done with this kind of waste?

One option, given that we no longer find it acceptable to leave it in the house, is to find a convenient piece of land to put it on. Another is to burn it. A third is to find some new use for it. How is it decided which of these options to pursue when faced with 'waste'?

Land spill or landfill?

Putting it on a convenient piece of land is a popular option. Anywhere from the remote islands of northern Scotland to the foothills of the Himalyas, you can see the consequences of people taking this route. The leafy lanes of Oxfordshire where I live suffer fly-tipping in common with many other communities – when running, I frequently encounter piles of discarded rubble interspersed with nappies and other household waste. Even the most supposedly romantic locations are not immune. Mountaineer Sir Chris Bonington has described the Everest base camp as 'the world's highest garbage dump' because so much litter is left there. There is even junk in space – an estimated five and a half thousand tonnes of it orbiting the earth, which by now would also be coated in a thin film of faeces from ejected space-person poo if early space-craft waste management had not been improved.

Of course most of us tut if we see crude piles of rubbish tipped anywhere, because in much of the UK (if we're talking about legal disposal as opposed to illegal) we've evolved 'dumping on land' to mean the sophisticated engineering that is modern landfill. The UK has an accident of geology that has made landfill a particularly favoured option. We have extensive deposits of clay and 'aggregates' that are useful to industry – think bricks, plates, road surfaces, gravel, etc. Having been dug out, they leave convenient holes to fill with rubbish. An alternative would be to fill them with water and turn them into nature parks – this has been done – but using them to bury rubbish often seems like a good arrangement. Two services from one hole in the ground.

In the early days of landfill the rubbish just went in the hole, and when it was full it was covered with earth and grassed over – end of story. And not just household rubbish – all sorts of industrial waste was mixed in. This was done using an officially sanctioned principle of 'dilute and disperse' – mix toxic waste with enough non-toxic waste and eventually it will all rot down and be rendered harmless, filtering out into the surrounding environment.

Eventually we realised that it might be a good idea to put more protection between the waste and the ground (particularly if there might be drinking water sources somewhere beneath that ground). The holes were lined, first with clay, and more recently with different layers of clay and plastic, so as to be impermeable. So now rather than diluting and dispersing, the waste and the considerable water that tends to collect in the hole, sits and festers to itself. In part, the wetting is welcomed because it helps decomposition of the large amounts of putrescible (biodegradable) waste such as food, grass clippings and paper. In areas of landfill that remain dry, you can excavate newspapers forty years old that are still readable. But the wetting also means the formation of noxious 'leachate' liquid from the percolation of water through the waste.

Since this is no longer allowed to filter through the bottom of the pits, it has to be removed and treated, or sometimes it is pumped back through the waste under a kind of 'flushing' principle to speed up the rate of decomposition. The decomposing mass also gives off gas. Other than when there were a few unfortunate incidents when the gas migrated to neighbouring houses and caused explosions, this has been managed by sinking pipes into the heap and taking the gas either to a flare to burn it off, or more recently by installing engines to generate electricity.

Landfill fans – and there are still some – insist that this is a controlled, safe and 'cost-effective' way of dealing with waste. Compared to just tipping it anywhere, that is of course true, but does it make it the right thing? I have been fortunate (in terms of seeing how they work) to visit many waste sites during my time serving on an advisory group to a large waste company. My introductory visit was to one of England's largest landfill sites, occupying worked-out clay pits in the 'brick country' area of Bedfordshire, on one of the hottest days of the summer. Thanks to the weather it was probably smellier than usual, but not shocking – there are legal conditions that landfills must not cause nuisance through undue 'odour'. What shocked was the sheer scale of the operation. Although only a little of what is tipped is visible at one time, because there are strict rules that require it to be trundled over by huge compacting machines and covered in soil (the 'daily cover' operation), what is visible displays the extraordinary diversity, in size and type, of materials and objects. Bin bags tend to get ripped as they are compacted (first in the waste truck and then on the tipping face) so that all the stuff we think we are secreting away and saying goodbye to, re-emerges under the wheels of the machines to briefly reprimand the watcher before disappearing under the next lot, or under the screen of dirt cover. All those nappies, wrappers, half-eaten ready meals, batteries, dead flowers. If you want to know what is in the average household bin, there are plenty of sources, but if you really want to know, go look at a British landfill.

But landfill is changing. Rules from the European Union are increasingly limiting the burying of biodegradable waste to reduce the methane emissions that come from landfill gas, because methane is twenty-five times more powerful than CO_2 as a contributor to global warming. So landfill in the UK will not be limited in future by lack of available space – there are still plenty of suitable holes in the ground – but by its potential to contribute to global as well as local problems. Following on from this lead, the governments of England, Scotland and Wales have adopted, or are in the process of adopting, long-term policies that seek to minimise waste to landfill, signalling that this is not the route of choice for the future.

The cleansing power of fire

The next on our rather short list of waste-management options is burning it. Here, the equivalent of the 'tip it anywhere you can get away with' option would be the backyard bonfire. Waste disposal by burning is an age-old practice, and it is only recently that science has told us that it is really not a good idea. The relatively low temperatures reached when burning waste in the open mean that some things do not burn completely, and 'incomplete combustion' produces dangerous chemicals. Standing over a 5 November bonfire, or indeed any bonfire, particularly one with plastic on it, means inhaling some very nasty pollutants, something not to be repeated on a regular basis. I tried not to mention any of this to neighbours when the first village bonfire party we attended featured a pile of pallets and hedge trimmings topped off with a beautifully balanced, orange, PVC-clad sofa.

Modern incineration, like modern landfill, has evolved (or rather has been regulated) into a relatively safe means of waste disposal, although controversy still rages around many new incinerator proposals. Many countries that have not had the UK's geological advantages have opted for wide-scale incineration to deal with their waste. Switzerland (mountainous), Holland (waterlogged) and Denmark (limited space) are among them. I can understand why the idea of burning waste and using it to generate power makes a great deal of sense to many people. Certainly compared to landfill, and even present-day materials recovery facilities (where our recyclables go for sorting), it seems cleaner, neater, more 'modern'. Everything goes one route, from a large, enclosed hall where humans rarely tread, into a giant iron-box furnace of almost stellar intensity. Some incinerators I have visited in continental Europe evince huge civic pride. They boast cutting-edge architecture and are platforms for public artworks; they provide electricity or heat, sometimes both, to grateful local communities, so I can understand the appeal. But I also understand the worries of those to whom incineration conjures dirty chimneys and quantities of noxious ash. Up to a third of what is consigned to the furnace comes out as ash. Some can be recycled into building materials, but if that is not possible it has to be taken to landfill, so objectors' instincts that incineration is not a complete solution are well-founded. In addition, the amount of energy recovered may not justify the destruction of potentially valuable materials, or compare well with other means of generating electricity – although recent technologies do perform better. Later, I'll come back to the circumstances in which incineration might be a good option for certain wastes.

Another life

So what of the third option, finding a new use for the waste? Of course, it wasn't so long ago that reusing things would have been the norm. Go back to the early years of the twentieth century in Europe and the household waste bin would have consisted mainly of ash and cinders from fires, and food waste, although the latter only if there weren't pigs, chickens or dogs to eat it. Paper would have been used as fuel in fires, glass bottles with deposits could be taken back and reused, and metals were used for scrap. Of course people had much less stuff anyway, and what they did have would have been so expensive relative to their incomes that it was in their interests to make things last or get them repaired. Durability and repairability were the norm, and whole businesses were based on making things usable again.

This 'make do and mend' state of affairs was maintained by the First World War, then a depression, and not so long after the Second World War. At that time re-use and recycling were regarded as essential – the Ministry of Supply compelled local authorities to mount 'salvage' programmes to ensure that valuable materials were not lost to the war effort. Rationing, and with it the impetus for careful use of resources, only came to an end in the late 1950s. But then economic growth and the pursuit of affluence became major political goals, and these required increased consumption. Some of the ingredients for souped-up shopping had appeared between the wars, such as 'department' stores and the emerging idea of shopping for pleasure, but it was only in the late 1950s and 60s that consumption really took off – society wanted to 'indulge itself'. Salvage started to fall by the wayside.

The quantity and composition of household waste started to change. Ash and cinders reduced as electricity and gas replaced coal fires for heating. At the same time there was an increase in packaging materials, made essential by a shift to self-service supermarkets rather than shops with assistants serving out goods. At the end of the First World War, British local authorities collected 9 million tonnes of household waste. By the end of the

Second World War it was 15 million tonnes, and by 2005 that figure had doubled to 30 million tonnes, while the population had increased by only 20%. The increase in weight was accompanied by a greater increase in volume as ashes gave way to packaging materials, such that some households began to need two bins for their waste, a new state of affairs. Greater volumes of food and garden waste started to appear, as well as innovations such as disposable nappies.

Fortunately for the local authorities left with handling this increase in discarded stuff, investment in landfill had begun before the Second World War, so disposal was sufficiently cheap and available not to provide a brake to increasing consumption. The stage was now set for a dramatic acceleration of what economist and philosopher Kenneth Boulding dubbed the 'linear economy' – make stuff, use stuff, throw stuff away. Putting stuff back into the system becomes a comparative rarity.

We need to think not just about household waste, but also industrial waste. Household waste has, since the mid-nineteenth century, been dealt with by public authorities because of the potential impact of accumulating wastes on public health. Industrial waste, however, is largely a private matter between companies producing it and other companies who can make money from getting rid of it. Where there's muck there's brass, as the saying goes. Non-household waste has always been produced in much larger tonnages than household waste – at present, in the UK, nine times as much. Industrial waste is the 'shadow' of the things we consume, a theme we'll return to.

The linear economy progressed apace through the boom years of the 1960s until the early 1970s, when there was a sudden bout of anxiety about rates of resource use. The 1970s saw the emergence of environmental groups, including Friends of the Earth, who saw recycling as central to a new approach to the environment, and chose as the focus of one of their first public campaigns the decision by Cadbury Schweppes to stop using returnable glass bottles. There was the 1972 'Limits to Growth' report, which warned of the possibility of resource scarcity and damaging levels of pollution – more about this in Chapter 8. Then there was the 1973 oil crisis, although the restriction of oil supply was not about absolute scarcity so much as Middle Eastern oil producers reacting to US support for Israel in the Yom Kippur War. Even so, it provided a taste of what a resource-constrained world might feel like. In 1975, the then Labour government announced a 'war on waste' and even set up a committee – the Waste Management Advisory Committee – to investigate how greater recycling might be secured. However, this effort foundered on the sheer economics of the problem – it was invariably more expensive to recycle waste than to bury or burn it. Why should that be?

One reason is that the reuse and recycling of materials was no longer a practical or financial necessity. New stuff was there to buy and people could afford it. If you are poor, there is an incentive to pick up waste and make it into something else, or sell it to someone else who can use it. To make recycling an attractive proposition economically, someone has to want the reclaimed materials – there has to be a market. Markets for 'recovered' or 'secondary' materials did exist at that time but were not consistent. The economy overall was by now much more geared for the use of 'virgin' materials.

The role of entropy

The second reason is entropy. Here we shift from economics to physics, and we need to get a grip on the underpinnings of our material world to really understand the problems of the linear economy.

Otherwise known as the second law of thermodynamics,

entropy describes the tendency of the physical environment to 'lose' available energy, and move progressively from ordered states to more chaotic, dissipated states. To reorder things requires energy to be put back. We experience this in everyday life – living in a house inevitably results in disorder, and to pick things up and put them back takes energy. It is only the fact that the earth has the sun to provide continual energy that allows complex biological organisation (i.e. us, and all plant and animal life) to exist in anything like a stable state.

As Kenneth Boulding puts it, nature deals with 'pollution' and 'waste' by recycling. A hefty injection of solar energy powers the various physical and chemical processes that put biodegradable materials back into the system – the carbon, nitrogen and phosphate cycles, and so on. We have moved outside those systems by utilising or inventing a wide variety of non-degradable materials. This explains why we have pollution problems and natural systems don't.

Entropy also helps to explain the economics of the industrial version of recycling. Much of the stuff we use to make things starts off relatively pure and discrete – the trees that become timber or paper, the oil that becomes plastics, even some of the metals that become myriad things. But then materials get mixed up into products, and have a host of things added to them. There are thousands of different types of plastic, and the additives used to give plastics different characteristics run into the thousands as well. Little bits of metal turn up in all sorts of things in different combinations. Even paper comes in lots of types and mixes – bleached, dyed, coated, laminated, corrugated – and glass in different weights and colours.

Once this plethora of complex products becomes waste the components are even more mixed up. This soup of stuff is of no use to anyone. To give the materials another useful life involves separating and purifying them again. For household waste, there are two ways this can done.

One is for the householder to separate the bits of material that are reasonably 'pure' – paper, cans, glass, clean plastic. These won't account for all the stuff thrown away, but at least it's a start. However, they must be collected separately and kept separate all the way to the reprocessing factory. The second way is to collect the mixed recyclable materials and use machinery and/or people to sort it out. The first option costs money to fund the collection, the second to fund the plant. Both tend to cost more than the alternatives, partly because of the energy needed. When natural systems recycle materials they make use of the free and bountiful energy of the sun. We are using largely fossil energy, which comes at a price, both financial and environmental. If there is no strong market to pay for the reclaimed materials to the extent that these costs are covered, it is not surprising that landfill or incineration will come out cheaper, as everything goes down one route. So from an energy point of view, reusing products is much better than all the bother involved in recycling them, because, strictly speaking, recycling means reprocessing, with all the resources needed to achieve that.

Supply push not demand pull

So if the economics is completely against it, why would anyone in the developed world bother to recycle? The answer lies in that dawning environmental consciousness of the 1970s and 80s. Back then the environmental agenda looked very different to that of today. The hole in the ozone layer had not been discovered. When I started work in the environment movement in the early 1980s, man-made global warming had been postulated but most people considered it implausible. The dominant environmental concerns were around visible pollution rather than these less tangible threats. Whole rivers rendered lifeless by industrial effluent. Forests killed and lakes unable to support fish because the smog from power stations and large industrial plants led to acid rain falling on them. Health concerns from landfill and incineration. Waste disposal of all kinds became a major worry.

Gradually, public authorities in many European countries decided that there should be an increase in recycling, as a better route for wastes than landfill or incineration. The UK came late to the party. Since the political efforts of the mid-1970s had not come to much, in 1983/4 recycling of household waste in the UK stood at just 0.8%. Industrial waste was probably doing better, but data is so poor that there are no figures to confirm the rate. But, following on from the more progressive European countries, and chivvied along by legislation passed in Brussels to restrict landfill, the UK's household recycling has leapt from 7.5% in 1996/7 to nearly 38% in 2008/9. No mean achievement, but still a lot less than countries such as Austria and Germany.

However, governments still had to solve the problem of finding markets. Initially, this wasn't easy. Although there are global markets for recovered paper, glass and metals, historically they have fluctuated unpredictably, subject to the amount of new, generally cheaper, materials available. Countries that started earlier with recycling, especially in continental Europe, had developed some markets to take them, although they freely admit that sometimes this involved paying people to take the materials rather than selling them, and often sending them long distances. It was a symptom of the difficulties that the UK government had in 2000 to set up a special agency, the Waste and Resources Action Programme (WRAP), to help to find markets for the materials collected. WRAP's work made an impact, but what really rescued the recycling effort in the UK was Chinese economic growth.

The Chinese industrial expansion was so fast, and its appetite for materials so great, that for the last decade or so it has been worthwhile for the Chinese to take our waste plastics and paper back on the boats that bring us their products. Without this market, we could be stuck with lots of stuff collected in the name of environmental concern, but unwanted in the UK or Europe. The frightening aspect of it is that the Chinese may lose their appetite for our stuff in the not too distant future as they begin to produce their own secondary materials. They have already said that they no longer want to import some types of plastic.

So what we have seen is largely 'supply push' for recycling, rather than 'demand pull', until the Chinese came along with genuine demand in a way that we can only hope lasts a bit longer. But in the main, materials are re-entering the economy not because the economy invariably wants them, but because the alternative means of disposal are being cut off. I wonder to what extent we would bother to collect and export waste plastic to China if we didn't have recycling targets to meet.

This is also true for a lot of industrial waste. A greater proportion of industrial waste is recycled, because it can find a market. However, because waste is a private matter between a company disposing of it and a company removing it, it will get recycled *only* if there is a ready market. Other than ensuring that waste is disposed of without causing harm to health, there is no public authority on hand to help the process of recycling along, whether by legislation or by subsidy. The really obstructive and scandalous thing is that we know very little about individual industrial waste streams and their recycling rates and potential for improvement, precisely because they have lain outside public interest and scrutiny.

But here's the rub. Why would we have an economy organised to recycle unless it is one populated by products that are designed to be recycled? We are a very long way from Evie and Ed's world where everything has been thought about from the perspective of what will happen to it at the end of its life. We have created the products first, and worried about their fate afterwards. Or rather, the people who have created the products have not been required to worry about their fate at all, and it is mainly local authorities that have had that burden thrust upon them.

And that takes us back to my mother's wheelie bins. Local authorities are accused of all kinds of incompetence, sometimes fairly, sometimes unfairly. On recycling, they do what they can on limited budgets to meet novel and fairly tough targets – and yes, to answer the question at the beginning of this chapter, what they have achieved does amount to progress. But it is progress with inbuilt limits. What really cannot be laid at the door of local authorities is the design of the products that end up in their multicoloured bins. That is a much bigger story: the story of the extraordinary diversity of stuff that makes up our lives, and makes up our overwhelmingly linear economy.

'Stuff' is products, but before it is products, it is raw materials. To get a grip on the implications of the linear economy we first need to understand a bit about those materials – what are they, can they be recycled, how much are they recycled? Why is it a problem if they're not? Because being linear means discarding products and materials from the economy before their useful lives are over. It also means wasting the other resources that have been used to make and transport those products, such as fuel, water and human labour. It is not just the *kind* of resources we use but the increasing *amount* taken that is the issue – the very high throughput of stuff in our ever-growing global economy. Some pioneering work on 'mass balance' (a way of calculating the material inputs and outputs of an economy), which was sponsored by waste company Biffa from 1998 to 2006, came up with the sobering figure that for every tonne of our personal consumption of stuff, ten tonnes of fuel and materials has been used, and 100 tonnes if including water. A 2009 Friends of the Earth Report estimated that for the industrialised nations overall, the amount of material consumption per year that we are each responsible for could be 15–35 tonnes, again without counting water. In the developing world it is less, but people there are fast playing catch-up, and why shouldn't they? So the amount of basic resource (fossil fuels, metals, minerals, timber and other crops) extracted from the environment has increased by 50% over the last thirty years and is projected to increase a further 40% in the next twenty years. That might not be so bad if the majority of those 'backstory' resources could be somehow recaptured, but the other extraordinary figure from the mass-balance analysis was that, of all the resources flowing into the UK economy (materials and energy), less than 2% was retained in the economy for more than six months, with the rest emerging as waste. Some of the materials would be recycled, but probably less than half. The rest is written off by the economy as having no value, but still has to be dealt with as waste, while the effort and environmental damage that went into acquiring it is also written off.

This wasteful consumption is putting unbearable strain on the life-support systems of the planet. We don't tend to see this because of the long journeys the stuff makes to get to us, the many hands it passes through, and prices that fail to reflect the high environmental costs of producing it. It is time to delve into the secret life of stuff.

Part 2
The Stuff of Life

4

From the very beginning

It is hard to know where to start with the stuff of life.
How we see the world depends a lot on our training,
values and disposition. If you are a chemist, you will
understand the world as a series of elements and mixtures
(compounds) of those elements, perhaps specialising in
'organic' chemistry (everything to do with carbon) or 'inor-
ganic' (everything else). The preoccupation with carbon
is not just because there are so many compounds
involving the very promiscuous carbon element, but also
because we are 'carbon-based' life forms. Could there be
other kinds? We may find out one day.

If you are an ecologist, you may see the world in terms
of 'renewable' and 'non-renewable' resources. Renewable
equates roughly with animal and vegetable: these things
have the capacity for reproduction and thus replenish-
ment. Non-renewables are the minerals, and they exist
in finite quantities. At the end of the day, everything we
use is either grown or mined.

If you consider yourself a consumer of 'quality' products, you may think you prefer the 'pure' to the 'mixed':
pure gold or other precious metals rather than cheaper
mixtures; and when it comes to fabrics, the 'natural' to
the 'synthetic': cotton, wool and silk rather than polyester and nylon. You may, however, be prepared to tolerate
Lycra in your knickers and sports gear, and stainless steel
– a comparatively recent metal mixture – in your cutlery
drawer. Apart from the fact that staying 'pure' is tricky

in today's mixed-up material world, the terms 'natural' and synthetic' can be fluid and misleading. Plenty of poisons are entirely 'natural'; plenty of 'man-made' chemicals are perfectly benign.

So where to start with the material world? Why not at the very beginning?

Elemental forces

Physicists tell us that the universe began with the 'Big Bang', a vast explosion of energy from which we are still feeling the reverberations. As the infant universe cooled down, energy became particles, particles became atoms, atoms became elements, and so began the material world. An element is a substance made of atoms of one kind only – once atoms of two or more different elements combine, they are compounds. Hydrogen was the first element to form, followed by helium. These two are still the most common elements in the universe, and they are what stars are mostly made from. Elements are distributed through the universe when stars explode, in a process that gave rise to planets like our own, and that will continue for aeons after we are around to think about it. A recent twist to the Big Bang theory is the idea that the universe may have been 'recycled' – contracting to a minute point and then exploding out again in an endless cycle. How appropriate, I thought, to the theme of this book.

Children's author Kate Petty wrote a wonderful book called *Earthly Treasures* about minerals and their uses. They are indeed treasures, but in a sense they are not just earthly, they are universal. We know that we share elements with other planets in our solar system, and probably with others way beyond. So those traces of metal in our bodies, where they play an essential role in our biochemistry, as well as in our mobile phones and other gadgets, have origins back through time scales that are not just geological but stellar.

Science's best guess is that life on earth formed some 4.5 billion years ago, through a chance combination of compounds, aided by energy from sunlight and possibly kick-started by lightning strikes. The earth was fortunate enough to have water, and it seems pretty certain that the ancient oceans were the cradle of our life – they enabled simple compounds of oxygen, hydrogen, nitrogen and carbon to be somehow sparked into the amino acids that we know as the basis of life. At some point, these proteins combined to form a molecule that could repli-cate itself, the precursor of DNA, the miraculous molecule that can unzip itself and make copies of each half and therefore reproduce. The rest is evolution: from single-celled organisms to the 'tree of life' familiar from every explanation of Darwinism. It branches early into animals and plants, having passed on from the fungi and bacteria, and from then on into myriad life forms. Humans sit – some may say precariously – on one of the many twigs of life's tree.

Darwinism is well known – there has been a huge media effort devoted to helping us to understand the theory of evolution and its political and social significance. When Crick and Watson cracked the genetic code by under-standing how different combinations of chemicals formed 'genes' and determined different characteristics of living things, our understanding of evolution was confirmed and enhanced. Now we can even rearrange those genes and create artificial forms of life, so some feel that our mastery is complete, even overweening.

Much less attention is given to the mothers and fathers of chemistry, whose insights are no less momen-tous. It started with Henning Brand, the seventeenth-century German whose quest for gold led him to boil up large quantities of urine and accidentally discover phosphorus, the first element to be isolated by a chemist.

Then there were Antoine and Marie Lavoisier, who started to understand how elements combine and separate by experimenting with the properties of oxygen. We owe a great deal of our understanding of the world to John Dalton, who came slightly later and worked out that elements are composed of atoms that do not change. It was then not long before the brilliant Russian Dmitri Mendeleev worked out how the elements relate to each other, and drew a map – the periodic table. The importance of this insight is hard to overstate, because it cracked the 'code' of chemical relationships in much the same way as Crick and Watson cracked the code of biological life.

Early in the nineteenth century, chemists realised that carbon is the key chemical of living things on our planet, and understanding of the carbon cycle got under way. Gradually, the earth's life-support systems (carbon cycle, nitrogen cycle, water cycle) were illuminated. The science of ecology was born in the twentieth century and drew heavily on insights from chemistry as much as from biology and botany. Ecology is the study of the interdependence of living things in complex systems – how a woodland works, or a rainforest, or a humble hedgerow. It goes beyond understanding the make-up of individual plants or animals or chemical compounds and looks at the world in a more joined-up way.

Recently, we have understood even more about the extent of the interdependence between the elemental, mineral world and the plant and then animal worlds that grew up on its sturdy foundations. Rather than the formation of minerals ending after the earth had formed, new ones evolved in response to plant and animal life – the calcite in the shells of sea creatures, for instance. Up to two-thirds of the more than 4,000 known types of minerals on earth can be linked in some way to biological activity.

Transformations

Human exploitation of natural resources began in the same way as that of other animals – eating what could be picked or caught by hand, finding places to shelter and building dens or nests. Where we departed from most of the rest of the animal kingdom was when we started to deploy 'technology'. What is meant by that? Technology is simply the application of knowledge for *practical* purposes, so technology is our body of knowledge about what works and what doesn't.

There are animals that make and use tools – some birds use sticks to poke out insects from nooks and crannies, apes use stones to crack nuts. But human ingenuity rapidly outpaced these innocent efforts, to produce sharpened stones for hunting, smoothed rocks for grinding grain and sun-baked clay pots for holding water. Fire gave us the ability to cook food, which may have provided the extra energy needed to build bigger brains. It also enabled us to do more interesting things with stuff dug out of the earth. We started to melt metals and re-form them to make better tools and ornaments. Fabrics made from woven plants replaced animals skins for clothing. Early experiments in chemistry involved using plant materials to dye fabric and minerals to make paint. We dub these early eras the Stone Age (lasting from 100,000 to 10,000 BC), Bronze Age (from 10,000 to 1,000 BC) and Iron Age (from 1,000 BC right up to AD 1,800, but these simple classifications hardly do justice to the variety and depth of early technical accomplishments. As historian Christopher Lloyd chronicles in his book *What on Earth Happened?*, if we look at the time since the earth was formed as represented by a twenty-four-hour day, humankind has had dominion for only the last three seconds – but what a busy three seconds those have been, and what a lot of stuff we have created.

Mankind's first big lifestyle change was to cease the nomadic life of the hunter-gatherer and settle down in villages, adapting wild plants into crops by selecting the best varieties to cultivate, and domesticating animals. This was in many ways a harder life, but it was what enabled settled 'civilisations' to establish. Farmers could produce surplus food for those who did not farm, which enabled an increase in populations. The freedom from searching for food also enabled other members of society to take on defined roles – leaders, merchants and warriors (it is very hard for a society without a food surplus and the ability to store it to establish an army). Some of the first really successful 'technologies' were employed in controlling water, for irrigation of crops and also to keep floods away from the otherwise very useful soils of river valleys. On the back of this success, civilisations in different parts of the world such as the Sumerians, Egyptians, Aztecs and Mayans had sufficient human and material resources to embark on the monumental building projects that left us the Pyramids and other spectacular monuments.

The Sumerians are credited with inventing the first wheels somewhere around the fifth century BC – not for transport first of all, although someone quickly thought of that, but to throw clay pots. At roughly the same time, skill with metalworking developed, from using 'native' metals such as gold and copper which are occasionally available and workable in their raw state, to mining 'ores', metals combined with other elements, and being able to separate, refine and shape these, smelting them in kilns originally developed to bake clays. Bronze was the first metal alloy – a mixture of copper and tin, which had superior strength to the individual metals, and was ideal for making weapons. A shortage of tin, caused by interruption of traditional trading routes sometime in the first millenium BC, may have led to the Greeks developing iron as a substitute for bronze, ushering in the Iron Age. Then, as now, necessity was the mother of invention.

Of all the ancient civilisations, it is the Romans we associate most with technological and material ingenuity. Their grand buildings, many still standing, were made possible by their development of an early form of concrete called *pozzolana* – volcanic earth mixed with lime. This, together with the innovation of using lead for pipework (lead came as a by-product of mining silver), enabled the construction of their complicated systems for trans-porting water – systems of aqueducts and canals running for hundreds of miles and ending in reservoirs, public baths and fountains. The Romans probably invented the daily bath.

In Britain, we tend to think of the time after the Romans retreated back to their homeland in the fifth century AD as the 'Dark Ages', although in reality trade and tech-nology continued to develop. Across the globe, particularly in the Middle East and Asia, material ingenuity continued apace. Then after the twelfth century, with the building of stronger ships, came an age of exploration – seafarers ventured to the corners of the known world, gradually realising that it wasn't flat, and filling in the gaps in the maps. Exotic goods – early desirable 'stuff' such as silk, jewellery and spices – were traded over huge distances. But the next big leap forward in material terms was the Industrial Revolution of the eighteenth and nineteenth centuries.

Britain is often credited as the cradle of the Industrial Revolution because of the invention of the steam engine in the late eighteenth century, but some historians see a shortage of wood as the major factor in accelerating mech-anisation. By the middle ages, much of the native forests of Europe, including England's, had been cleared to make way for agriculture, to provide cooking fuel, and to build houses and ships. War was particularly expensive on trees, and after the English Civil War in the mid-seventeenth century, the exploitation of Britain's abundant coal began in earnest. Coal is a more powerful source of energy than wood – a tonne of coal is equivalent to two tonnes of dry wood. More available power meant more effective machines; this in turn meant new processes and products, so this explosion of industrial energy in part kick-started the Industrial Revolution. Mechanised looms for textiles, paper mills, steel mills, chemical plants, railways – they are all phenomena of the last two hundred or so years. Of greatest significance for building was the use of steel. The blast furnace had come from China to Europe around the turn of the sixteenth century, leading to the widespread use of cast iron, but it was the invention of the Bessemer process in 1855 that led to mass-produced steel and this enabled the construction of the great bridges, railway terminals and civic buildings of the nineteenth century.

Also in the nineteenth century, mankind started to exploit oil. This triggered another big lifestyle change, because oil has been the biggest factor in our increased mobility. Coal and steam were all very well and gave us trains that are now treasured heritage objects, but without oil and therefore petroleum to feed the newfangled internal combustion engines, and then later diesel (for trains), and kerosene (for aircraft), there would not be quite the choice of means of getting about. There would also be no plastics. Oil in the form of bitumen seeping naturally from rocks had been known to mankind for some centuries, but it was only when the American George Bissel thought of boring deep into the ground to look for oil in the 1850s (he'd seen it done for salt) that the oil industry really took off. Some histories credit a Mr Edwin Drake with first striking oil in large quantities in Pennsylvania in 1859, but others claim that Azerbaijan did it nearly a decade earlier.

The next 'age' could be thought of as the 'polymer age' with the coming of plastics, but in most histories it has been eclipsed by another element, the most plentiful on earth – silicon. Even if we would like to think of ourselves as living in the 'information age', somehow leaving physical materials behind, without the silicon to make the chips there would be no computer power. Without the chips we could not have gone into space, even though the computers used to take *Apollo* 11 to the moon had a fraction of the processing power of today's mobile phones or even pocket calculators. Computing power has also enabled the expansion of our understanding of physics. The current focus is on the Large Hadron Collider, a device to smash particles together in the hope of learning more about their constituent parts. Without modern computers, this exercise would not be feasible.

So here we are, a species that has found ways of utilising just about everything in the earth's crust, has mastery over the foundations of biological life to the extent that we can manipulate the genetic code, has seen the earth from space, and has peered into the sub-sub-parts of atoms. It is hard not to think of ourselves as at the pinnacle of a process of evolution, top of the food chain, and masters of all we survey. We were perhaps temporarily humbled by the first pictures of the earth rising over a lunar landscape taken by *Apollo* 8 astronauts in 1968: the achingly beautiful sphere of blue and green, the sense of isolation and the fragility of the life-support systems, unique as far as we know. But for most people life has gone on without carrying a burden of fear about our position on planet earth. The natural world as inherited from our cave-dwelling ancestors has been radically changed, but to each of us in our present personal worlds, little of that change is visible – much has disappeared more or less noiselessly – and as a whole the world is still considered very much at our disposal. What does that mean for our relationship with stuff now?

5

Deconstruct your home

How much do you know about the materials that make up your waste, indeed support your life before becoming waste? You may be familiar with the arguments about packaging – whether plastic is better than paper, light-weight better than recyclable, cartons better than tin cans. You may be a parent agonising about reusable or dispos-able nappies. You may know the best thing to do with dead batteries. But are you really 'materials literate'? Do you know enough about where it all comes from, and why it is as it is? Without better 'materials literacy' it is hard to understand why the linear economy is such a problem, and what we might do about it.

Imagine you decide to undertake a complete inventory of all your stuff and sort it as far as possible into types of materials. You might have to start by turning out the entire contents of your dwelling into your garden, if you have one, and include all the garden furniture and accou-trements. You should include all means of transport such as cars and bikes, and any buildings. If you were being really thorough you'd demolish them. Don't forget the clothes you stand up in and the various gadgets and trin-kets that are likely to adorn your person. Don't really feel up to doing that? I'm going to do it for you, albeit on paper.

The inspiration for this exercise comes from several sources. One is from America's oldest environmental organisation, the Sierra Club, whose book *Material World* took 'statistically average' families from a range of countries in five continents and photographed them with all their worldly goods in front of their houses. In the pictures from the US, UK and other Western countries, the smiling families are accompanied by all the expected paraphernalia, looking slightly uncomfortable with all their sofas and lamp stands and general clutter in the middle of the street. In other countries, the possessions are to our eyes pitifully few, although the subjects are still smiling, as people tend to do when facing a camera. The subjects were asked what was their most treasured possession – for some it was animals; for others, pots, blankets, Bibles, pianos. Only one family refused to play that game and said, 'Nothing – our children.' But as well as examining what individual possessions mean to their owners, the book gives a graphic account of the vast disparities in material wealth that endure across the globe.

The second source of inspiration is a television programme popular in 2002/3 called *Life Laundry*, in which the friends and relations of diehard hoarders brought in a hit squad in an attempt to de-clutter them, before the mess resulted in estrangement or worse. The hit squad consisted of a psychologist and an antiques/ recycling expert, one to persuade the offenders to part with their stuff, and the other to find useful places for it to go. The programme began with making the subjects turn all their stuff outside, where it often filled a tennis court or similarly large space. Admittedly, these people were chosen because they were bad cases, but the sheer volume of stuff was arresting. What could it all be? That is indeed the important question: the idea is that until people see stuff divorced from its surroundings, they can't make good decisions about its purpose (or lack thereof).

The third source of inspiration is the artist Michael Landy, who occupied a vacant store in London's Oxford

Street in 2001 and set about publicly dismantling and destroying all his material possessions. This was done in a supremely controlled, almost clinical way: a conveyor belt wound its way around the store carrying trays of Landy's stuff, while he stood on a gantry above selecting items for deconstruction by a team of assistants. Everything was meticulously inventoried and nothing escaped being broken down into granules of matter – his own artworks, gifts, clothes, photographs, letters, his car, his pet's toys. Seven thousand, two hundred and twenty-seven items were reduced to 5.7 tonnes of landfill. As a performance artwork, it was an invitation to those watching to confront the material reality of possessions and understand what they are made of, as the granules of once important items were sorted into sacks of material types. But it also invited them to consider what their own possessions meant to them – people shopping in Oxford Street could wander in to watch things similar to the things they had just bought being systematically destroyed. Some were shocked by the abandonment of sentimentality and the denial of attachment to personal possessions; others found it 'cleansing'. For Landy, it was like witnessing his own death – after all, this process of sorting and disposal is what happens when we no longer need our material possessions. And yet Landy describes this as the happiest two weeks of his life, a liberation from the process of consuming. Afterwards, although he had to reacquire many things in order to resume a 'normal' life, he found himself maintaining a fairly lean material existence, acutely aware of the superfluity of many things that previously populated his life.

In all these cases, leaving aside for the moment the complex psychology of possession, what is in front of us is a motley collection of materials. They have been taken from their usual places in homes, ostensibly fulfilling some useful function, and left naked and shivering on public view. What are they and where have they come from? But more important, why do they matter to us as a species, rather than simply to us as individuals?

Not that many kinds of material

So we have taken apart your abode and its contents, and the car, and thrown in the garden furniture and tools. Then we have sorted the materials into broad categories (some of them still as intact objects – we don't have Michael Landy's army of 'deconstructers' at our disposal to break everything down into tiny bits). What do we have?

- A large heap of **minerals** – most of it brick and concrete, but including **glass** and a lot of **metals** in all sorts of guises
- A smaller but still sizeable heap of **timber**
- A surprisingly large heap of **paper**
- A very colourful heap of **plastics**
- A larger and equally colourful heap of **textiles**

Not many different piles, although lots of different things within them. But what we will notice from the piles is that materials don't often fly solo, as it were – there are many things made of more than one material, to the extent that, for some things, we may have difficulty assigning them to a particular pile. A fridge for instance – is that mainly metal or plastic? A mobile phone? Which pile should have the sofa? It uses wood, plastic (in the foam, unless it is old enough to be stuffed with horse-hair), metal (for strength) and cloth. It is a mixed-up material world, which is a big factor in the linear economy, as we shall see.

And that's without considering the less visible mixtures – the supporting cast of innumerable chemicals and other additives that expand the possibilities of materials almost beyond imagination. These are the inks, glues, preservatives, finishes, lubricants, fillers, solders, dyes, conducting substances and a host of others, that enable combinations of metal and plastic to become mobile phones or televisions, or natural fibres to become washable suits or magazine paper. They include the detergents, bleaches, brighteners and pleasant smells that cleanse houses and clothes, and the chemicals that guard our gardening efforts or keep the ants out. They are everywhere and anywhere, and represent decades of technological specialisation and refinement. Whole industries are devoted to making our paper whiter as well as our washing, and the label on our bottle of fizzy drink as brightly coloured as possible. These additives are crucial to the secret life of stuff.

But before we get into the complexities of products, we need to get back to basics – the materials, their origins, their uses, and their destinations.

6

Exploring the material world

Minerals

We started by saying that all our materials are either grown or mined. Let's start with the mined.

Nature's cycles use very few minerals – most of them are present in small amounts where they perform supporting roles, like the calcium and magnesium that strengthen bones, the chromium that helps to regulate our blood sugar, or the iron that is essential to photosynthesis in plants and oxygen transport in us. Humans, however, have prodded and poked the earth's crust until it yielded up over ninety elements, combined into hundreds of useful mineral compounds.

Minerals are the backbone of the economy, stiffening its structures and propelling its progress. Without metals, a large and important category of minerals, there would be no reinforcements for big structures such as bridges, few machines, none of the forms of transport we take for granted, no conducting of electricity into our homes, no electronic devices, and much less packaging capable of protecting food for months on end, whether tins or metal-lined drinks cartons. There would be few weapons – blunt sticks only – and no surgical instruments. Without the other mined and quarried minerals, there would be no concrete for buildings, no bricks for houses, no tiles for roofs, no cement. Without oil and coal, the 'fossil fuels' that we count as minerals, there would be a fraction of the energy available that we have been used to, and there would be no plastics or synthetic fabrics, because they come from oil. Without minerals life would be lower-rise, slower, and with very few moving parts. It would also be very much less well fed – the phosphate fertiliser we need for food production is a mineral resource. In short, modern life would be impossible.

Life was, however, without many of these things for the large part of human history. Materials guru Michael Ashby identifies 1800 as the point at which the developed world crossed the threshold from being *mainly* dependent on renewable resources (timber and other crops, plus animal by-products) to being mainly dependent on non-renewable, mineral resources. Now, we are almost totally dependent on minerals for everything, including food production.

Concrete

Metals are not the only important minerals. A colleague once said to me, as I was holding forth about plastics, 'If you really want to worry about something, worry about concrete.' Mankind's largest consumption by far, if looking at global tonnages of stuff, is of construction minerals, particularly the 'ceramics' – minerals that we have transformed through heat. The foremost of these is cement, which is made from calcium (from baked limestone), silicon (from sand or clay), and small amounts of aluminium and iron. Thanks to having to bake rock to temperatures in excess of 1400°C, cement-making is the most energy-hungry process after electricity generation, and also releases CO_2 as a consequence of the chemical reaction that results in the hardening of the material. Concrete is a further step, obtained by mixing cement with sand and gravel. At least ten billion tonnes of concrete are produced globally a year using nearly three billion tonnes of cement, with the total increasing 4% annually. Bricks also come into the category of 'ceramics' since they are baked clay.

The advantages of these construction materials are enormous. They are versatile and their raw materials are widely available. Brick, concrete and cement can make structures of all sizes and shapes that work in any climate. However, as well as the energy needed for their manufacture, there are drawbacks that exemplify the linear economy. Stone is relatively easily recyclable, in the sense that a chunk of stone from one dwelling can be taken for use in another. The house where I spend summer holidays in Scotland, for instance, may contain stones from a nearby Iron Age 'broch' (a kind of fort occupied by a number of families), meaning that those stones may first have been shaped by people some 2,000 years ago. If buildings are dismantled carefully, bricks can be reused. But the way in which concrete and cement have been transformed by heat means that they can't be easily reused in chunks, they can only be crushed up and used as fill materials in roads, and to a limited extent fed back into the concrete-making process. Concrete is also considerably less durable than brick or stone. The maximum life of one of today's concrete buildings depends on how well built and maintained it is, or whether it keeps up with architectural taste, but it is likely to be in the tens rather than hundreds of years. Contrast that with the Roman brick-built buildings still standing after 2,000 years, or the Neolithic stone homes in the Orkneys, some of which are largely intact after 4,500 years.

So modern building materials take lots of energy to make and are hard to recycle. What of their raw materials? Rocks and bits of rock (the latter are known collectively as 'aggregates', which includes sand, gravel and various types of clay) are plentiful, particularly in the UK thanks to our geology. So running out is unlikely. The disadvantages of using these materials are more a result of the environmental disruption wrought by obtaining them. Extracting them means either dredging them up from the seabed, or breaking up large quantities of rock by blasting. Done on a small scale with picks and shovels, as our early ancestors would have done, taking rock and aggregates would not have a huge impact. Done on a massive scale in huge quarries, as it is in parts of the UK and other countries, it can scar landscapes and destroy wildlife.

There is also the human impact of quarrying. It generates traffic, sometimes many truckloads a day, it is noisy and it releases large quantities of dust. Nothing galvanises neighbourhood cohesion like the threat of a quarry, as I discovered shortly after moving to my area of Oxfordshire, much of it lying on potentially profitable gravel beds. To add insult to injury for those living near them, empty quarries can mean convenient holes to fill with rubbish – landfill.

Pottery

Before we move away from brick and concrete to metals, we mustn't forget that included in the pile will be all our household china and glassware.

Baking clay into pots goes back to the Stone Age, some 25–50,000 years ago. Simple sun-baked, and later fire-baked, clay pots would have been porous, so no good for holding liquid, where animal skins would have done a better job. It took another 10,000 years to invent glazing, literally an application of glass by coating the pot in a fine layer of sand, which kiln-hardens to a waterproof, glossy and hard-wearing film. This rendered the pots watertight, making them more useful and longer lasting. Today we use the term 'china' because, as with many other technologies, the Chinese refined their methods of using clay to produce the most delicate and beautiful examples of the craft. They also had extensive deposits of white 'china clay', which makes porcelain, infinitely superior to the rough red clay we know as 'earthenware' or 'terracotta' and the stolid grey version we call 'stoneware'.

Pottery overall represents quite a small proportion of the stuff we use, and a small proportion of the use of aggregates as raw materials. It is relatively durable if handled carefully, although today's cheaply produced dinner sets come at a price that suggests ready disposability even if they don't break first. But being 'ceramic', i.e. heat-transformed like bricks and concrete, pottery is not recyclable – the only things you can do with discarded pottery are to use it in place of aggregate in concrete, or as a dry powder filler in some plastics and resins. In the linear economy these are not common uses. When my husband suggested a 'china shy' for our village fete I was horrified, but a quick tour of local charity shops flushed out boxes and boxes of donated but unsold china whose only destination would have been landfill, so letting the kids smash it up first wasn't as terrible as it seemed. Perhaps we need to revive the days of treasured china being passed down through generations.

Glass

Glass, on the other hand, is much more helpfully recyclable than china. Although it is also a 'ceramic', it has a more fluid molecular structure – oddly, it is technically closer to being a liquid than a solid. This explains its readiness to be shaped, as well as re-melted and reshaped.

Even so, glass has an artisan, treasured face and a linear economy face. For the first, think of Venetian beads, delicate blown glass bowls and goblets, crystal heirlooms. For the second, think of the smashed beer bottle on the street, the discarded light bulb, the jam jar not pressed back into use but lobbed a quarter-full of mouldy jam into the wheelie bin.

Glass-making goes back probably 4,000–5,000 years. For something that can be so exquisite, glass has very humble ingredients. Most of it is silica sand, plus other minerals including soda ash, dolomite and limestone. These bits of rock are heated until they melt and fuse together, for a few minutes yielding something malleable as the glass-blower or machine work magic and turns the humble materials into something useful or beautiful or both. The thick liquid rapidly hardens when the heat is turned off, and the new form is set. I have watched glass-blowers at work in Venice (where they have been exiled to the outer island of Murano because they kept setting fire to the main city), a craft of extraordinary beauty and skill, passed down through generations. At the other end of the scale, commercial glass production features huge smelters and rollers and moulding machines, and a river of molten tin on which to float the glass if making flat glass. It is a process not unlike a steel mill in its fiery intensity.

Each of us in the UK consumes an average of 331 glass containers (bottles and jars) per year, and that's without the glass in windows or less obvious things like televisions and computer screens. The majority is made here, but we also import a lot in the shape of bottled wine and beer.

The advantages of glass lie in the cheapness and ubiquity of its raw ingredients, as well as its resistance to water, corrosion and attack by chemicals. You can store the most vicious acids in glass as well as the finest wines. Their durability makes glass containers highly suitable for reuse, as they were in the days of deposits on glass bottles. In an economy where beer and wine come from all over the world, this is no longer regarded as practical, although countries such as Sweden and Denmark have managed to maintain successful deposit-refund systems.

With care, glass should be reusable indefinitely, and in theory it is also recyclable indefinitely. In practice, the contaminants that come with collected glass (the bits of paper, plastic, corks, metal lids, etc.) create unhelpful air bubbles or lumps when the glass is being recycled, diminishing its quality. Keeping colours separate (which is important for the end uses of recycled glass) is not as much of a problem as it was since automatic sorting technology can separate the clear, green and amber. But if you do put glass out for recycling it is helpful if it is clean and if the colours are separated as far as possible.

However, finding the end uses for different coloured glass is where it gets tricky. About 1.6 billion wine bottles a year are imported into the UK, and these are mainly green glass. In the other direction, we export a lot of spirits, particularly whisky, bottled in clear glass. So thanks to the conscientious recycling of the wine drinkers, we have an enormous amount of green glass 'recyclate' but we still have to manufacture clear (flint) glass for our exports. To help to overcome this, the Waste and Resources Action Programme (WRAP), the government agency set up to help establish stronger markets for recycled materials, is trying to encourage more bulk importation so that wine arrives in 24,000-litre 'flexi-tanks' made of plastic, and is then bottled in the UK in lightweight bottles from green glass recycled here. Every 24,000-litre tank saves shipping 32,000 wine bottles around the world. The linear economy has got a little bit less linear.

Overall, however, in the UK we only manage to recycle around 60% of glass used in packaging, and less than half overall, and landfill is still the final destination for the majority of the rest. For one thing, we have a much lower density of bottle banks – one per 2,700 people – than most other EU countries, where the best have one for every 1,500 people. Collection direct from households varies around the country. On top of that, some of the biggest users of glass bottles such as pubs and restaurants, still send huge loads crashing into landfill.

Recycling glass saves energy. Glass-making, like its near relative cement-making, is very energy-hungry. If a glass furnace uses 50% recycled 'cullet' and 50% new raw materials, it uses 15% less energy compared to using 100% new. It also uses less water and saves raw materials. The sand used to make glass is plentiful but still requires something being dug up with all the associated environmental and human impacts. So if extensive reuse of glass bottles can't be achieved, recycling glass is incontrovertibly worthwhile, so long as it is turned back into the bottles and jars it used to be, rather than being used instead of crushed rock in roads – a recently developed habit.

So our linear economy is responsible for eating away the landscape and the seabed as well as using large amounts of energy in pursuit of the materials that form a large part of the fabric of our lives. We have evolved an extraordinary global dependence on a few humble raw materials but, sadly, ones that cannot always be easily recycled once their work is done. In the UK, every single one of us 'consumes' four tonnes of aggregates a year, and getting on for five tonnes if you add in cement and some other minerals. And this is without the impacts of the next important category of minerals: metals.

Metals

We saw in chapter 4 that human exploitation of metals has defined whole eras. Over the millennia we have moved well beyond copper, bronze and iron and have continually added to the repertoire of metals that we have been able to extract from the earth's crust and press into service. Seventy of the ninety-two 'naturally occurring' elements are metals, and they are all in use in some form or other. It is impossible to visualise modern life without the benefits of these diverse and versatile resources.

Our use of metals is greatest for the 'workhorse' metals, those existing in large quantities and used for many applications. These include aluminium and iron, the third and fourth most common elements in the earth's crust after oxygen and silicon. At the other end of the scale there are those of relative scarcity and much higher value – gold, silver and platinum. We revere gold because it is one of the most chemically unreactive elements of all – it never loses its shine, forming a currency that can be passed down through generations. Many metals are used in very specialist applications, invisible to the casual consumer. Metals like scandium, used to give extra strength to aluminium in high-quality bicycle frames. Or yttrium, a silvery metal named after a village in Sweden, and invaluable for superconductors and lasers. Then there is niobium, whose corrosion resistance and ability to take on beautiful colours make it the ideal metal for body piercings. There are the 'rare-earth' metals, recently come to the fore as crucial to new 'green' technologies. Some metals we have used for millennia but now demonise for their toxic properties, like lead, mercury and cadmium. Some metals have the potential to end humanity's reign on the planet – uranium and plutonium, the stuff of atomic bombs.

These valuable resources are scattered unevenly across the globe. The 'developed' world, having been through an industrial revolution earlier than other places, has largely mined out its easily accessible reserves and has shifted its attention to less developed countries. The impacts of mining are increased by the fact that most metals do not come 'pure' like some deposits of gold, silver and copper – they exist as 'ore', that is, minerals that contain a percentage of metals, but are combined with other elements such as oxygen and sulphur, so a process of refinement has to take place.

Mining can be done in a variety of ways. When it comes to environmental and human impacts, all types of mining can be done well or done badly, regardless of the size of operation. At one end of the size scale there is what is called 'artisanal' mining, which means individuals or small communities breaking rocks with picks and shovels, often illegally and under very harsh conditions, to get what ores they can out of the ground. There are an estimated 100 million people supported in this way in developing countries. Dubbed 'subsistence mining', it can mean essential income for impoverished people, but the health cost is often high. Artisanal gold miners in Africa and South America use shocking quantities of highly toxic mercury to extract gold, for instance, despite better techniques being available if they were trained to use them. Then there is deep mining, which is the kind probably most familiar to those of us in the UK, where coal but also metals such as tin and copper have been mined for centuries by digging deep shafts and then tunnelling sideways. There is 'strip' mining, which is for ores that are not too far beneath the surface, so that a layer of earth can be stripped off, the ores removed and the earth replaced. Then there is open-cast mining, deeper than taking off strips, but not as expensive as tunnelling, and which involves simply digging a large hole.

Mining itself doesn't take up huge areas of land. Compared to other 'primary' industries such as forestry and agriculture, its local impacts are relatively small, but they can be serious. Mining creates different types of waste – the 'overburden' of soil and rock that has to be removed to get at the ore; the waste rock which does not have enough mineral content to be of interest; and then 'tailings', a slurry of ground-up ore, still containing the chemicals used to extract the ore, and as much as 70% water. A common problem from the waste rock is 'acid drainage' – rocks left on the surface can contain sulphur compounds, which when exposed to rain produce sulphuric acid, causing catastrophic water pollution if there is no water treatment in place to deal with it. Heap-leaching involves extracting metals such as gold and copper by percolating arsenic or sulphuric acid through the crushed ore, and this process can contaminate soil and water. Other pollution problems come from traces of the ores themselves left behind in waste rock (which is many times the bulk of the useful ores, sometimes a thousand-fold or more), since metals such as lead, cadmium, arsenic are toxic if they get into soil or water in sufficient quantities from inappropriate dumping. Good practice in mining now forbids dumping in rivers, but it used to be common. Tailings dams have been known to fail, drowning people in poisonous mud, so they have to be carefully constructed and maintained. The worst loss of life to date was from the failure of tailings dams associated with a fluorite mine in Italy in 1985, which killed 269 people.

Mines can also have more far-reaching effects – the worst-case scenario is that they open up to traffic areas that might have been previously untouched, so where mining companies go, logging or clearance of land for agriculture can follow. The Trans-Amazonian Highway in Brazil is one of the worst examples of this, built to supposedly clear land for previously landless people, but actually opening up large areas of the forest to mineral exploitation. Then there is what happens when the mine is worked-out and the valuable minerals exhausted. A well-restored mining area can be an asset, as the Eden Project's 101 things to do with a hole in the ground attests – Eden is itself in a former china clay quarry. In developed countries, restoration is often a matter of regulation or accepted good practice; it is embodied in the ICMM Sustainable Development Principles, for example, which the major mining companies have all signed up to. Unfortunately, it is not unknown for mining companies to delay or walk away from meeting their clean-up and restoration responsibilities.

Mining can bring considerable benefits to communities in terms of wealth and development, but only if earnings are equitably distributed. Often the people bearing the health and environmental costs are not those reaping the benefits of the materials. In the UK we tend to see little of this as we have long since exhausted most economically-viable supplies of metals (the tin and copper in Cornwall for instance), and for most people it is only the disturbance and dangers of coal mining that are within living memory. As with many other commodities, we have 'outsourced' the impacts of mining to less developed countries, where environmental controls on mining may be poor, and health and safety non-existent.

These impacts are before we even consider all the energy and water needed to extract ores, and then, having extracted them, purifying the metals (smelting) and turning them into useful objects. These impacts mean that it is essential to make the best use of the metals already in circulation, and minimise 'primary extraction' as far as possible. Metals, like glass, can be infinitely re-melted and recycled, and this uses much, much less energy than starting from scratch – in the case of aluminium about 95% less. Fortunately, although metals are a non-renewable resource, they are also a very durable resource, and even when scattered through the economy they are still largely there to recover if we design things properly.

Aluminium is a good case study of metals in the linear economy because it is in so many things around us (drinks cans, ready-meal containers, kitchen foil, cars, computers, window frames, door handles, etc.). Its ubiquity is because it is one of the lightest, most flexible and low-toxicity metals (unless ingested in high concentrations), making it particularly suited to food packaging. It is used not just for drinks cans and foils, but also in very thin layers in cardboard cartons (where it helps to keep liquids from going off) and crisp packets (where it helps to keep the contents crisp). The main raw material of aluminium is bauxite ore; it is plentiful, and comes from places as far flung as Australia, China, Russia and South Africa. The ore tends to exist near the surface, so is mined by scraping away the soil. It is then shipped to a processing plant, milled into a fine powder, boiled up with caustic soda, and the crystals of aluminium oxide that form after a few days can then be smelted. Smelting (melting of the oxide and getting the pure aluminium to separate out) takes so much power that aluminium has been dubbed 'congealed energy.' Fred Pearce's book, *Confessions of an Eco Sinner*, says that it accounts for 2% of world electricity consumption, and he paints a vivid picture of the inferno-like conditions in an Australian aluminium smelter, a single plant that uses as much electricity as a city of a million people. And this assumes that the aluminium is smelted where it is mined. Often it travels great distances to countries with cheap hydroelectric power like Canada and Scandinavia before being transported back to South-East Asia to be made into a product.

Recycling can mitigate at least some of these impacts. We consume nearly 900,000 tonnes of aluminium in the UK every year, split between building, packaging, transport and engineering uses. Packaging is only a fifth of that consumption, but is the most visible, and drinks cans alone account for 90,000 tonnes That averages out at nearly 150 cans a year each. Technology has enabled great efficiencies in the use of the metal: cans can now be made from a sheet of aluminium thinner than a human hair, and even taking 5% off the weight of a can translates into a saving of 15,000 tonnes of aluminium across the EU every year. But how successful are we at recovering this valuable material?

Not as successful as we should be. The high cost of production means that scrap metal of all kinds has value, so if anything from bashed-up cars to empty drinks cans are collected and bulked up they will find a ready market. Unfortunately, that value is not high enough to overcome some of the barriers to recapture. Globally, only 30% of aluminium production is from recovered sources. In the UK, we still only recycle 42% of aluminium packaging overall and and 55% of aluminium cans. These losses are for the very simple reason that people don't always put cans in the recycling bin (assuming they have one) or because they are discarded outside the home and recycling in public places in the UK is still very under-developed. In addition, there is nothing on a can to convey the story of the huge human and environmental cost of providing that simple object, and the importance of keeping the resource in circulation.

The human race has mined the earth, ceaselessly, fearlessly and messily for many generations. Its products have been dispersed all over the globe, and while a good proportion of the main metals are recovered, overall, we are continuing to mine and produce new supplies rather than shifting the emphasis to recovery, despite the eminent recoverability of metals if used with sensitive design. Meanwhile, demand is going up – more than for any other global resource – as all the developing countries of the world, one by one, undergo massive industrial revolutions. But the pain is still being borne by poor people and remote habitats all over the world, in many cases without proper recompense.

Wood

Trees are one of nature's most remarkable organisms. Technically, a tree is a plant with a main trunk and a distinct 'crown' of branches, but that dry definition hardly conveys the majesty of California giant redwoods, or the history embedded in a thousand-year-old yew tree in an English churchyard. Sadly for trees, it is usually their trunks that are of most interest to us commercially – for fuel, timber and more recently to 'lock up' the carbon that trees take from the carbon dioxide in the atmosphere.

When we think of wood, particularly solid wood, we tend to think of good furniture, perhaps antiques, and maybe high-quality toys. We mentally contrast the warm, natural, enduring qualities of wood to the hard, synthetic, characteristics of plastic, even though much furniture now that calls itself wood is made from veneers (very thin films of wood) and resin-bonded woodchip, and is not designed or intended to be any more lasting than most of the plastic we use.

In past generations, however, the significance of wood was principally as a fuel and as a construction material. If we left the UK's landscape to do what it wanted, it would revert almost entirely to woodland dominated by oak and ash (what is called a 'climax' ecosystem). This is how it was before human populations grew sufficiently to eat into it, and eventually clear it nearly completely. In tropical regions the 'climax' ecosystem is also often forest, although of different kinds, including the rainforest that we call 'jungle'. Then across vast swathes of Canada, Scandinavia and Russia there is the 'boreal' or 'northern' forest, composed of conifers, whose resin-coated needles can withstand the cold. These are the main types of forest, although the UN Environment Programme distinguishes no less than twenty-six types.

The forests of England were almost entirely gone by the eleventh century. Globally, more than half of the original forest is gone, and a further quarter may be gone before 2050. For countries that still have them, forests now have to play a variety of roles. They provide food, fuel and medicines for local communities, but increasingly they are also part of a global trade in timber, some of which gets turned into wood products, and some of which is pulped into paper. Globally, half of felled timber is still used for fuel, less than half for sawn timber (for buildings and furniture) and around a tenth for paper.

That timber trade has a number of guises: sensitive, insensitive, legal, illegal. At its best it extracts a limited number of full-grown trees with least possible damage to the surrounding forest, allowing younger trees to grow on, new trees to be planted in gaps, and the rest of the ecosystem to carry on relatively undisturbed. At its worst, logging is carried out illegally on land that should belong to local people, involves the 'clear cutting' of whole swathes of forest, destroys the environment and local livelihoods, pollutes rivers and minimises the chances of the forest regenerating itself. According to the Forest Disclosure Project, illegal logging and over-extraction is occurring in some seventy countries, and accounts for around half of all traded timber production in tropical forests. There are, of course, many shades of practice in between the best and the worst.

This can make the environmentally responsible buying of wood (and indeed paper) a minefield. Most of the 18 million cubic metres of wood used in the UK is softwood: quick-growing trees such as spruce, pine and birch, grown in Europe, and a lot of it certified as having been grown 'sustainably'. There are several European countries where the overall growth of forest exceeds what is taken every year, so that the level of use of the timber is not a concern. On the other hand, there is still an appetite for exotic hardwoods from the rainforests for products such as garden furniture and kitchen worktops. Many of these come from trees that take decades, sometimes centuries to grow to maturity in natural forests and so have built up dense, durable grains of huge value in furniture and building – mahogany, teak, ramin, iroko. Ken Fim's book *My Journey with a Remarkable Tree* charts the journey of a tree whose wood is known in the UK as Keruing. It is felled illegally in the forests of Cambodia, where it has spiritual and social significance for the local people, smuggled into Vietnam, and ends up in a garden centre in England. From the sublime to the suburban.

There are schemes for certifying 'sustainable forestry', like FSC (Forest Stewardship Council) and Programme for the Endorsement of Forest Certification (PEFC), that aim to help buyers discriminate between 'good' and 'bad' wood. As we will see later, these schemes are important but not comprehensive.

The linear economy, at least in the UK, fails to recover much wood. To be fair, wood is not always easy to reuse and recycle, although most council waste sites welcome it. It tends to rot; it may be embedded in buildings and getting it out damages it beyond reuse; or it may have been heavily treated with preservatives, paints, varnishes or other forms of protection. The main way of recycling wood is to chip it or pulp it, and reconstitute it using resins and glues. That is how we get chipboard, beloved of budget furniture-makers, and MDF. MDF is made from woodchips that have been broken down to their constituent fibres through heat and mechanical pulping, then mixed with resin and wax. This goo is then hot-pressed to form fibre boards, the mesh of fibres providing a better tensile strength than would have been possible with chips and sawdust alone, although with all the pulping, hot-pressing and drying, the process is fairly energy-demanding. These materials are themselves hard to recycle because of the additives, and some commentators have concluded that by the recycling stage it is better to revert to the habits of our ancestors and realise the value in wood by burning it for energy. But overall, given the extent of deforestation worldwide (an issue we'll return to), the imperative is to be more careful about how we use trees and consider the pros and cons of alternative materials.

Paper

Most of the prodigious amount of paper we use now comes from trees, whereas in the past it was made from a great variety of other things. The beautiful vellum of previous centuries was made from animal skins, and indeed still is for specialist uses such as the printed versions of British Acts of Parliament, because the paper has a life expectancy of five hundred years. The ancient Egyptians' papyrus was made from the reeds in the Nile; the Chinese made paper from rice husks, and it can even be made from animal dung. The Eden Project shop has a nice line in 'ele-poo' paper and in Tasmania they are making it out of wombat droppings. In fact, paper can be made from numerous sources of cellulose fibre (the substance that surrounds plants cells and gives plants their flexibility) without going anywhere near a tree.

In past times a lot of paper was made from waste cloth, thus giving the plants that yielded the cloth a double life. I once visited a traditional water-powered paper mill in Basle in Switzerland, where huge wooden hammers pounded torn-up linen rags in water until they were reduced to a pulp, a process that had been going on since the sixteenth century. Men in leather aprons held large screens of mesh under the surface and lifted them deftly to catch a layer of pulp and let the water drain away, and in a miraculously short time the pulp was dry and a sheet of paper separated from the mesh. They allowed us tourists to try it – we managed rather less elegant results, but I still have my sheet of handmade cloth paper with the insignia of the mill stamped confidently at the top.

One reason why wood fibres are so suitable is that they can produce paper of very high quality and strength, but with a very low weight and bulk. Producing thin paper for large books would be very difficult from cotton or textiles, so trees now dominate. Modern papermaking on an industrial scale is a far cry from the water-powered hammers and chaps in aprons of Basle. The timber is brought in, stripped of its bark (which can then be burned as a fuel), chopped into chips, and then either ground up into a mash by huge grinders similar to millstones (mechanical pulping), or stewed in chemicals (chemical pulping) to separate the fibres. As well as large quantities of chemicals (according to one estimate 270 kg to every tonne of paper) all this involves massive quantities of water, which is why paper mills tended traditionally to be sited on rivers or lakes. Modern paper mills often feature internal 'kidneys' to clean the water and allow it to be recirculated many times, to the extent that some benchmark mills have an almost closed water loop with near-zero water consumption and exemplary pollution control. They can also capture and reuse the chemicals. As with many other industrial processes, however, this may not always be the case in less developed countries.

The slurry of fibre, by this time 99% water, is sprayed onto a moving continuous loop of mesh, allowing the water to fall through and be recirculated, and seconds later depositing the wet sheet of paper with a set of huge rollers. These squeeze out more water, after which something recognisably paper is moving through the machinery. What happens after that is determined by the paper's final use, which could be anything from humble newsprint, boxes for your burger and fries, to high-specification medical applications that require special coatings. At the end of the processing, rolls fit for a giant's till or toilet are taken off to be cut, printed, bound, or employed for a myriad other uses.

In the UK, some paper is still made from timber grown in Scotland, but most 'virgin' (i.e. new) paper pulp comes from Finland or Sweden, two of the major pulp-producing countries alongside Canada, USA and China, to be turned into finished paper here. Most of the wood pulp reaching our shores is grown in dedicated plantations as a crop, or taken from countries where annual timber growth is bigger than the amount taken. The notion of 'destroying rain-forests to produce paper' no longer applies to paper in the UK market, although a very small amount comes from 'old-growth' or 'original' forest – in other words, forest that has been there for aeons before humans found a use for it. Old-growth boreal forest in Canada is still being extensively logged, for instance. Such forest is irreplaceable – although trees can be replanted, the delicate interwoven webs of plant and animal life that have evolved with them are gone forever. A small amount of what reaches the UK as paper or pulp may have come from countries like Indonesia and Vietnam, and a further small proportion of that might have been logged very badly or completely illegally. For an eye-opening account of the *worst* aspects of the trade, including the destruction of precious forests in Canada, Russia, China and Vietnam, you need to read Mandy Haggith's book *Paper Trails*. A forest activist for many years, Mandy has witnessed the legal and illegal pillage of forests for paper, and chronicled the extraordinary lack of control of this primary resource in some countries. Despite the efforts of environmental groups and some enlightened sections of the paper industry, there is still habitat destruction to feed the world's paper habit.

Worldwide, demand for paper is still growing – it is projected to increase by 77% from 1995 levels by 2020. In the developed world, including in the UK, paper consumption is dropping as a consequence of moves to electronic media, as well as the 2009 recession, when falling advertising revenues led to a contraction in the size of newspapers and magazines. But that is from a high base – we as a country are the fifth largest consumer of paper in the world. Our consumption of magazines, for instance, is prodigious: according to the Periodical Publishers Association there are 3,200 consumer titles in the UK, consumers will spend £2.5bn on magazines in 2010, and they reach 87% of the adult population. What's worse is that a sizeable proportion – as much as a third to a half

– never even leave the newsagent's shelf, because the cover price is so high and the cost of paper so low that it is better to overstock than risk running out and missing a sale. A good proportion of these are sent to recycling, but all the bits and pieces on the covers used as enticement to buy (CDs, packets of seeds, even flip-flops) are likely to end up, in pristine condition, in landfill.

Overall, there is an estimated 12–13 million tonnes of paper and cardboard a year in the UK going into books, newspapers, magazines, packaging, stationery, junk mail, receipts and all the other bits – that is equivalent to over twenty billion copies of this book. We are surrounded by so much paper that it is not surprising that it has become a symbol of waste. In developing countries, however, they could do with more paper – for education, communications and packaging to stop needless waste of food.

Fortunately, paper can be recycled. Recycling uses far less energy than processing virgin pulp by mechanical pulping, which needs a lot of electricity (which if made by burning fossil fuels will have a high carbon footprint). On the other hand, recycling sometimes uses more energy than chemical pulping, because a by-product of the chemical process is a liquid that makes a very good fuel which can be used to power the pulp mill. As incentives for bio-derived energy increase, some paper companies are contemplating going more into the energy business, farming trees as a 'biomass' crop, and producing pulp, paper and fuels on a single site. There is also the 'de-inking' waste from pulp recycling to deal with, which is another chemical soup that needs careful handling, although again it is sometimes used as a source of fuel.

Paper can't be recycled forever – after six or seven times of going through the process, the fibres weaken and are no longer any good. This means that a complete 'closed loop' system, where paper is continually turned back into paper without cutting down any more trees, is impossible. It is estimated that if only waste paper was available, paper production would last for six months, so input of 'virgin' fibre will always be needed. That is even more reason to ensure that the sources of the virgin fibre are carefully vetted. There is also a paper hierarchy – whiter-than-white premium-quality paper can only be made with about 17% recovered fibre, whereas tissue can be more than 60% and newsprint and packaging more than 80%. Since most of what we put out in our recycling bins is newspaper and magazines, that's what it is suitable to be recycled into. Here is a relative success story – thanks to the previous UK government encouraging a 'voluntary agreement' with the newspaper industry, newsprint is now on average 80% recycled, with some mills using 100% recycled stock.

How much of the paper we use ends up being recycled? From the household, around a third. From offices and industries, it is harder to say because the figures are incomplete, but overall (counting households as well, and accounting for imports and exports) the recovery rate is nearly 70%. A lot of this goes to China to be made back into paper, since despite the newsprint success, in the UK we have a limited number of mills to reprocess paper, and cannot provide a market for all the waste paper collected. At the same time, those paper recyclers that do exist in the UK are complaining about the increasing number of local authorities collecting 'commingled' or 'mixed' recyclables from householders because this has resulted in waste paper being more contaminated with other materials, which increases their costs. This is getting more serious as the downturn in paper use has a knock-on effect on the amount put out for recycling, meaning a looming shortage of good-quality paper to put back into the recycling process.

Taking into account all the paper that ends up in archives or bookshelves of one kind or another, there are still several million tonnes of mashed-up tree that could go round the system another few times, ending up in landfill every year. From the household, this will generally be because people still don't put paper into their recycling bins. But from businesses, it is because there is no incentive, either regulatory or financial, to sort, store and organise a separate collection for paper.

So our linear economy treats paper not as a treasured, crafted luxury as in past times, but as a mass-produced commodity of immediate disposability, from the free newspapers that litter the streets of every town, to the cups that hold our coffee, to the paper tissues we blow our noses on. Similarly timber – we make furniture designed to last for years, but we also make disposable chopsticks, dispensable packing crates and pallets, and cheap, ephemeral furniture. This might be all right in the parts of the world where timber is a 'crop' and can be harvested and processed in ways that don't do environmental and social damage. But where this consumption is contributing to the destruction of old and precious forests, and where a resource that is highly energy-intensive to produce is used once and then wasted, it is not.

Plastic

Plastics – we love them and we hate them. The very word means flexible and adaptable, and we celebrate those properties by using plastics in a million ways. But we have also come to associate them with litter, mess and an inconvenient durability once their immediate purpose is served. The ubiquitous plastic bag is a particularly bad ambassador, lying around in ditches and residing as tattered streamers in trees. Bits of plastics have even formed enormous, swirling, intractable masses in mid-ocean. They have usurped their role as material servant and started to feel like a master.

Part of the inspiration for this book came from stumbling across an exhibition entitled '100 Years of Making Plastics' at London's Science Museum, where it struck me that there are few opportunities to really study the history and properties of materials in an accessible way. This is particularly true for plastics, despite what amounts to a rich history, a huge diversity of applications and some fairly difficult chemistry.

Plastics are polymers – big molecules in the form of chains or networks, which is what gives them their helpful qualities of strength and adaptability. Polymers exist in nature – rubber, amber, and the cellulose and sugars in plants are all examples – but generally when we think of polymers we think of substances coaxed into existence by chemists. The plastics timeline starts in 1907 with Bakelite, the first 'true' plastic in that it was the first to be made entirely from synthetic materials. Belgian chemist Leo Baekeland mixed together phenol (otherwise known as carbolic acid and made from oil or coal), formaldehyde (which is made from methanol, a type of alcohol distilled from wood) and ground-up wood particles known as wood flour. He applied carefully controlled heat and pressure and ended up with a hard, mouldable plastic, which he dubbed Bakelite. The new material found uses in everything from radio and telephone casings to billiard balls and buttons, most of them a comforting, warm conker brown, but some more brightly coloured. The house in which I grew up, a 1930s semi in Pinner, still had its Bakelite door-handle plates. For fans of retro, some very beautiful photographs of Bakelite objects can be found at the Ghent Virtual Bakelite Museum.

Not long after Bakelite, which now feels like a rather distinguished, pedigree kind of plastic, came polyvinyl chloride (PVC or just vinyl), made partly from oil and partly from the chlorine in salt. As the twentieth century progressed, more and more of these miraculous materials populated the industrial designer's armoury – polystyrene, polyester and many other polys.

Most plastic is a 'co-product' of oil. That means that by refining oil we can get petrol and other fuel products, plus the 'feedstock' for plastic, a chemical called naphtha, without going to much extra trouble. If we didn't make petrol, we might not bother with plastic, or at least not oil-based plastic – we would make it out of plants. At the end of the day, the chemistry is common, because oil is dead, squashed plants and animals (carbon), and living plants just as happily yield carbon. In the future world of Evie and Ed, that is precisely what is happening, because the oil era has come to an end.

The use of a non-renewable resource is not the only reason to rethink plastics – doing the chemistry on the naphtha to produce plastic is itself very energy-intensive. There are also concerns about the toxicity of some plastics, both in the production process and the effects of some of the additives used – we'll return to this when we look at toxicity later.

But at the present time, we split oil and make polymers – the UK has large oil-refining and chemical industries churning out the stuff. In the UK we make here about half of the 5 million tonnes of plastics consumed here every year, for more applications than anyone could possibly list. Just in one train journey, for instance, I notice plastic in the carriage bodies, the seats, the fabrics, the teacup and the little containers of milk, in bags, mobile phones, laptops, shoes, buttons, carrier bags, magazine wrappers, etc. etc. According to the trade body Plastics Europe, plastic consumption worldwide has shown extraordinary growth over the last sixty years – up from 1.5 million tonnes in 1950 to 245 million tonnes in 2008, with a slight recession-caused dip in the last couple of years.

Although there are many variants of plastic polymers to do different jobs, most fall into a few main categories of what are termed the 'commodity polymers' including polyethylene (PE), PVC, polypropylene (PP) and polyethylene terephthalate (PET). Globally, consumption of each of these is greater than any metal apart from steel. Packaging is the largest single use of plastics in tonnage terms in Europe, and packaging is what is very visible to us as consumers. A lot of debates about plastics become synonymous with the debates about packaging, even though plastics are much more widely distributed through our lives. Without packaging there would undoubtedly be more food waste before it even got to the consumer, and this is something the world can ill-afford, particularly as supply chains get longer and food is brought greater distances. Tonne for tonne, wasting food means wasting far more resources in terms of the energy, water and environmental disruption involved in producing it than could be accounted for by wasting the packaging. In addition, plastics are held up as a good packaging option because of their lightness, flexibility and strength. None of that means however, that we shouldn't consider carefully how the impact of plastic materials can be reduced.

At the moment in the UK we recover less than a quarter of the 5 million tonnes of plastics used every year, despite the fact that the majority are recyclable. We also have to be careful about the term 'recovery' – in some figures, recovery means recycling *plus* recovering the heat energy in plastics (they are, after all, a type of oil) by burning them in incinerators, also termed energy-from-waste plants. But on the basis that it is generally more efficient in terms of energy saved to recycle rather than to incinerate, let's consider why that figure is so low compared to other materials. Those little numbers in a triangle of arrows chasing each other on packaging help to illustrate how recycling ought to be easier, but it has been made harder by the way we use the materials.

In case it helps, here's what the numbers mean. 1 is PET – often clear plastic, what most drinks bottles are made of, but also appears as fibre (fleeces) and even credit cards. 2 and 4 are types of PE – high density (HDPE) and low density (LDPE) respectively. HDPE is often coloured or cloudy rigid plastic, so used in milk cartons and coloured bottles like those used for washing liquids, and things like beer crates where strength is needed. LDPE is used for less demanding applications like carrier bags and films. 3 is PVC, which can be rigid or flexible depending on the additives, and is used for everything from water pipes to fake leather fabrics. 5 is PP, very similar to PE and used in a similarly wide range of things, including margarine tubs and ice-cream containers. 6 is PS, a more brittle plastic, what CD cases and yogurt pots are made of. 7 is the unhelpful category of 'other', which could embrace a huge range of plastics with different properties, but most of our uses of plastic, particularly for packaging, fall into one of the other six categories.

This relatively small number of widely used materials should mean that plastic is not much harder to collect and recycle than different grades of paper. Polymers are actually very expensive – at the time of writing running at $1,000 per tonne, more than any metal except steel because of the high price of oil. But turning them into products is relatively cheap because they are easily moulded and also a tonne of polymer goes a long way. What further skews the economics of recycling is the relative lightness of plastic. For local authorities, who have to recycle a certain proportion of household waste by weight, the easiest thing is to collect paper to meet their targets – they would have to collect a lot more plastic than paper or glass to make a significant contribution to that target. The good news is that as soon as householders are offered the service of having a range of recyclable materials collected from the doorstep, they tend to use that service to full advantage and put lots of plastics into it, so that local authorities have begun to find it worthwhile to collect them. The Chinese buy most of our mixed plastics (i.e. other than bottles) because they have a shortage of polymers and can afford the labour to sort them into the different types, although UK markets for reprocessing of plastics are gradually developing as food manufacturers look for packaging solutions closer to home and mechanical sorting technology improves.

But then the next hurdle can be that products or packaging using mixed plastics or mixed materials are difficult to separate into their constituent polymers. To recycle successfully, reprocessing facilities have to be able to get a clean stream of specific polymers (such as PET) so that they can be chopped up and then purified by chemical processing back into that polymer – mixed polymers are much less useful and valuable. The more a 'closed loop' is aimed for (turning a waste plastic back into the same product) the more important is this purity. So for instance PVC labels on PET bottles are bad news because just one can contaminate a whole batch of PET to an unacceptable level and it will be rejected. I saw this in action at the Closed Loop Recycling plant in Dagenham. Because sorting technology is not foolproof, piles of bottles sent to the plant for recycling might actually end up in landfill instead. What Closed Loop owner Chris Dow really wants is the use of such labels to be 'designed out', but there are no standards that require this. So the little numbers in the triangles really only speak of the physical properties of each polymer, not of the way the products are designed, and certainly not of the economics of collecting them.

Plastics are immensely useful materials, but they also feel like the pinnacle of achievement for the linear economy, and an example of entropy (that inevitable dissipation of materials and energy) doing its worst. Most plastics are designed and priced to be quickly discarded, even though many of them are extraordinarily durable and perfectly recyclable.

Plastics often come in little bits of stuff, or break up into little bits of stuff, but they do not for the most part break down completely in the environment, i.e. 'biodegrade' in a way that allows their molecules to return to nature's cycles. In 1997, the oceanographer Charles Moore, while returning from a sailing competition, identified a mass of small pieces of plastic hundreds of miles wide floating in the Pacific Ocean. The bits had been borne on ocean currents from the discards of countries all over the globe, carried by the wind, washed down rivers or off beaches or chucked off boats. The sea had rendered them a uniform size of a few centimetres with the result that they form a layer both under the surface of the water and on the seabed. According to environmentalists, the patch forms a net that catches and accumulates pollutants, damaging marine life, and the disintegrating bits of plastic are mistaken for plankton by grazing fish, concentrating the pollutants in their bodies.

Even if we stopped plastic production tomorrow, some of the plastics already in the environment could persist for upwards of four centuries. In developing countries, all materials, including plastics, often have enough value for people to gather at least some of them back into economic use, but even here the recovery is not comprehensive and the use of plastic means that more traditional, degradable and possibly locally appropriate materials get supplanted along the way. We need to learn to see scrap plastic in the same way as scrap metal – a material that is resource-intensive to produce and therefore should not be allowed to dissipate into the environment and be lost from use. At the same time, we need to work out how to shift from using a non-renewable resource to using renewable ones – more of bio-based plastics later on.

Textiles

Fabric. We all have our favourites and our pet hates. Cotton is 'natural'. Wool is warmth. Cashmere is luxury. Silk is exotic. Some materials have definite places in time – Crimplene belongs to the 1970s and few people would sing its praises these days, although I feel bound to admit that in 1973 a pair of turquoise Crimplene hot pants were the pinnacle of cool in my class, especially if paired with a crumpled cotton cheesecloth top. Viscose sounds synthetic and a bit plasticky but is actually made from wood. We think we understand fabrics, but do we?

Humans have been dressing themselves for at least 50,000 years. As people wandered northwards from their African cradle, animal skins offered protection in colder climates. It took at least another 20,000 years to invent the first needles so that skins could be joined together – what a boon that must have been. But even before that, there is evidence that humans realised that twisting vines and animal sinews produced stronger cords for making tools and they even decorated them with beads made from pebbles and teeth. It was then a relatively short step to weaving, first with plant fibres from trees and grasses, later with domesticated crops. Then people realised that animal hair could be spun – sheep, alpaca, llama depending on the part of the world. By 6,500 years ago, there were different kinds of looms and a large variety of dyes derived from clays, roots and bark. The urge to dress up and decorate hit us very early on.

Today, the large-scale sources of fibre for textiles (not just clothing, but furnishings) are relatively few. Global fibre production divides pretty much fifty–fifty into the natural and the synthetic. Natural fibres are dominated by cotton, the product of the fluffy seed heads of *Gossypium*, an unprepossessing plant native to the Americas, Africa and India. Other traditional natural fibres are produced on a much smaller scale – flax (linen), hemp, even nettles. Wool, once the basis of great wealth and sometimes injustice in large parts of the UK, has a steadily declining market as synthetic fabrics that are just as warm but easier to care for have come on stream. Silk, spun from the cocoons of moths, has maintained its niche as a luxury fabric from ancient times to the present day.

These natural fibres were supplemented in the twentieth century with man-made versions. The invention of synthetic fibres is the same story as the invention of plastic – they are also polymers derived from oil, processed to be flexible and capable of being spun into fabrics. The earliest was nylon, followed shortly after by polyester. Today, global synthetic fibre production is dominated by polyester. Another version of synthetic fibres comes not from oil but from wood. Viscose, Modal and 'Tencel' (the trade name for lyocell) are produced by pulping wood and releasing its natural polymers in a process like papermaking. These 'regenerated' fibres can then be spun into fabrics with a silky quality.

Different fibres have different pros and cons. Natural fibres are in theory 'renewable', but that doesn't mean that there aren't problems with their cultivation. Cotton is one of agriculture's thirstiest and most pest-prone crops. Three-quarters of the global cotton crop has to be grown with irrigation, using staggering quantities of water – 2,700 litres per T-shirt – sometimes in regions that can't really afford the use of that water. Cotton is grown in the middle of the Californian desert with water 'mined' from an aquifer, for instance. And although cotton accounts for less than 3% of the area of world agriculture, it accounts for a quarter of global insecticide use. Wool has the advantage that it could be seen as a co-product of meat, but only if land is not overgrazed and degraded. Synthetic fibres are cheaper to produce, but the downside here is a large energy use. They are also, just like most other forms of plastic, non-degradable. Producing the raw fibres is also only part of the story – they have to be processed, dyed, spun, woven, finished and cut. All these stages have environmental consequences, from carbon emissions to local pollution.

Textiles is one of the areas where mixing of materials has resulted in huge innovation. A large proportion of textiles contain mixed fibres, carefully blended to offer the right combination of properties. Pure cotton is coveted by many people, but it is more prone to shrinkage and creasing than if mixed with synthetic fibres; similarly wool. This mixing is a problem for recycling (if by recycling we mean reprocessing back into what it was before). Like paper or plastics, both natural fibres and oil-derived polymers can be recycled, but only if they can be separated to yield a sufficiently 'pure' stream, and that requires expert, and therefore expensive, sorting. An increasing vogue for cheap trimmings of all kinds means that recyclers have to deal with ribbons, rhinestones and even rocks attached to fabrics. Dyes and finishes also get in the way. Fibres such as wool, when they were more widely available and less mixed with other fibres, used to be recycled as 'shoddy', which despite the current connotations of the word made perfectly good garments: the label 'pure new wool' was to distinguish the new from the recycled, indicating that it wasn't otherwise easy to tell. The only substantial recycling that goes on now is to produce mixed fibres for the unseen padding in mattresses, car doors and dashboards.

But perhaps more than the mixed-up fibres, it is our behaviour when consuming textiles that exemplifies the linear economy. Our consumption in the UK is over 2 million tonnes a year, divided roughly sixty–forty into clothing and other textiles (such as carpet and curtains), and 2 million tonnes of textile waste is generated each year. An insurance company survey suggested that the *average* UK woman has fourteen garments that have not been worn in the last year, and possibly have never been worn at all. When finally discarded, many of these must go straight in the bin and not even make it to a charity shop, since only around a third of the clothing discarded every year is recovered for resale or recycling. Of the carpet and other kinds of textiles, figures are not good, but much less is recovered. So nearly half a million tonnes of clothes and other textiles is buried in landfill annually along with half a million tonnes of carpet, including as much as 8,000 tonnes laid for events and exhibitions for just hours or days. The recycling rate for domestic carpet is a feeble 0.5% annually.

We expect to change our clothing and our furnishings on a regular basis whether they are worn out or not, a luxury of modern times. Previous generations with scarcer resources would create patchwork quilts and also mats out of strips of worn-out clothes and linens, creating colourful, multitextured 'rag rugs' to warm cold stone or splintery wooden floors. The 'newest' ones had pride of place in the parlour. When a little worn, they went to the bedroom, then to the kitchen, then to the scullery, and eventually to the dog. Nothing escaped the house unless it was completely worn out. Imagine that now.

Composite futures

We've got to the end of the list of materials, but what this leaves out is the growing number of clever 'composite' materials, mixtures of plastics with metals with ceramics with fibres, and so on. One of the earliest 'composites' was 'vulcanised' rubber, invented in 1839 by Charles Goodyear. By mixing the natural polymer (the sap from the tropical rubber plant) with sulphur, Goodyear made a product that was stronger and more durable, but in doing so ensured that it never bio-degrades. Rubber tyres used to be disposed of to landfill, but once this practice was stopped in 2006, there was a danger of mountains of used tyres piling up. These are a fire risk – one dump in England smouldered for thirteen years. Work by WRAP helped to establish new uses and markets including increased re-treading (so that the life of the tyre is doubled) using rubber crumb for playground safety surfaces, and burning the tyres as fuel in cement kilns. It took a concerted effort to stop that particular composite material becoming a major problem.

More recent composites have been developed to combine lightness with strength, so glass fibres or carbon fibres embedded in plastic are used for applications such as lightweight aircraft, car bodies and super-strength bicycle frames. Carbon fibre in particular is growing as a material choice. Made by cooking strands of plastic without oxygen until they 'carbonise', it produces a fibre that can withstand a very high mechanical load, and when combined with plastic makes a material that is mouldable and very tough. However, as Michael Ashby points out, the very thing that creates their useful properties, the hybridisation of two very different materials, makes them nearly impossible to recycle. So far, the options are to grind them up and mix them with new resin (although that produces a material that might be regarded as inferior in quality to the original), burn off the polymer fraction to leave the carbon fibres behind for reuse, or extract the fibres with special 'supercritical' fluids. None of these recovers the materials completely.

If the material is designed to have a long life, such as in an aircraft, the energy benefits of its lightness might outweigh the inability to recycle. Some have drawn attention to a dilemma over the use of composite materials in wind-turbine blades. The stress on the blades demands materials of very high strength, but with the large planned expansion of UK wind power, some estimates suggest that by 2040 there will be 380,000 tonnes of worn-out blades to dispose of, and as yet there is no obvious way to reclaim the materials. This might be one of those trade-offs worth making, but for more readily disposable applications such as household appliances, there is more of a concern. The high-tech qualities of composites also means they can be eclipsed in the marketplace by the next generation of even more high-tech materials, making it hard to know what kind of materials we might be dealing with in ten years' time.

A lengthening shadow

Let us return to our virtual inventory of stuff, sitting outside our virtual house. I have tried to give you a quick tour of where the material components of your stuff come from, a tour that embraces vast distances of space and time. The artefacts you live with every day have their origins going back generations, sometimes aeons, in human history, and have been derived from every corner of the globe.

We also know that all these materials will at some point end up as waste – our waste. Of the piles we have, how much is recyclable? First, of course, you should ask how much is reusable, i.e. could be used intact by someone else rather than broken down, since that will save energy. Of the construction materials, there could be roof tiles, whole bricks, windows and doors. After those, probably just about everything is reusable; at least 90%, I'd say. (I'm using % by weight here, which is very crude, but it's hard to know what other basis to use.) Your car, bike, furniture and fittings, clothes, gadgets, kitchen equipment, books, magazines, and hobby bits and pieces could all be found suitable homes through eBay, charity shops and the freecycle website. As to actually recycling, if reuse is not an option, it is not very easy to say – it rather depends on what kind of stuff you have. But we could take a few guesses. Anything metal, if it isn't in tiny bits or bonded inextricably to something else, can be recycled and has high value, so probably more than 90% of your pile – including steel beams, clapped-out kitchen appliances, gadgets, etc. The construction materials that are not re-usable are probably only recyclable if they can go to a crusher, and any wood from construction would probably have to be burned because it would be too contaminated or damaged, so overall for the house materials maybe 50–60%. For the paper, easily 80% would be recyclable, if not contaminated. Plastics and textiles, much harder – maybe only 30–40%. So your virtual pile of unavoidable landfill is still going to be quite large.

And, lest we feel that this is not too dismal a picture, this inventory of domestic discards gives us a clue only to the materials we use ourselves. It says nothing about what was involved in getting them to this point. Behind each newspaper, consumer product or piece of packaging, there is a whole 'shadow' of waste about which we know surprisingly little. This is the 'industrial waste' of materials, but also of energy and water, which is entailed in bringing us stuff. In the UK, non-household waste (classified as the totality of 'commercial', 'industrial', 'construction and demolition', 'mining' and 'agricultural' wastes) outweighs our household waste by a factor of 9:1. But in terms of thinking about the waste generated in pursuit of consumer goods, a lot of that won't even be in the UK statistics because it will have been generated abroad. We outsource not just our manufacturing, but our wastefulness as well.

Granted, we recycle more industrial waste than household in the UK (because, as we noted earlier, it is generally more homogenous and occurs at a larger scale, so is easier to deal with on an 'economic' basis). But good recycling rates are not a feature of all the countries from whom we import, assuming they even know their rates. So we are part of a global linear economy sending staggering amounts of stuff annually to an early grave, somewhere, anywhere.

The real problem we face in trying to get to grips with this situation is that we don't know the detail of those flows. We have basic information on material flows around the world, but the really useful data on how much of what goes into which products and how much is discarded and then recycled somewhere else, is largely missing. The priority we have given to economic growth, and lack of attention to the basic resources that fuel that growth, mean that we have only the crudest idea of the fluxes in the material basis of the economy. This obfuscation of how we use resources could undermine our well-being in just as dangerous and far-reaching a way as the obfuscation of the money men in the credit crunch of 2008.

Next, we will examine *why* it so important to know about what comes from where.

Part 3
Why Stuff Matters

7

Nothing comes free

'We are consuming resources as if we had two or three planets, but when I last looked we only had one,' declared Hilary Benn, then Secretary of State for the Environment, from a conference podium in late 2009. Spare us the glib remarks, I thought, and tell us what that really means. Which bits of this planet-sized problem do we need to worry about most?

In the previous section, I attempted to give a sketch of the pros and cons of the materials that make up the stuff we use. Nothing we use is without environmental impact of some kind. Humans have been appropriating and modifying the natural world for their own ends for at least 50,000 years. The ability to do so is for some people what marks us out as human rather than simply animal. This ability is also the reason that we have been able – at least to a certain extent – to defend ourselves against the assaults of nature: weather, disease and natural disasters.

What gives us cause for concern now is the extent of that appropriation, and, of course, how long it can go on rising as global population rises. Scientists and historians tell us that we are now in an era dubbed the 'Anthropocene' – a geological epoch all of its own, in which humans have identifiably altered the planet's basic life-support systems. To decide how bad this is and what needs to change, we need a picture of how all those materials, products and subsequent wastes, with all their varied implications, add up to spell either disaster or deliverance.

Let's think first about the things that constitute what is called 'natural capital', the physical underpinnings of the economy. This includes land and the natural resources and ecosystems that the land supports – everything from metal and timber to the cycling of water and nutrients. Why should this vital 'natural capital' be under threat? The chief underlying problem of our consumption is that the effects on the environment of making stuff are largely unpriced. If you are a mining or logging company, you have to pay people to work the mines or cut down the trees, but you don't have to pay the environment for the damage caused unless a landowner or a government has found a way to secure recompense. Even then, that money doesn't necessarily pay for repair, and no amount of money can replace lost species or habitats. The same goes for the rivers polluted by factories, or the health effects of dirty air. In a world driven by prices, for the environment to get a look-in, either the prices must be adjusted (to discourage some forms of exploitation and encourage others) or the damage must be taken out of the control of the market and made subject to regulation. This already happens in the developed world to quite a large extent, and increasingly in the developing world – but not enough. How this is best done in future we'll come back to in the final chapter.

For some resources, there is simply no alternative. We can't make more land. Although some countries have done quite well reclaiming land from the sea, it is a relatively small amount for tremendous effort, and land at or below sea level will always be vulnerable – think of the Netherlands, the Maldives, Florida. Land can be cleared of its native plants and animals to grow crops or to build on, but eventually there will be no more to clear. And by then we will have had a catastrophic effect on another non-substitutable 'resource' – our fellow species on earth. Once gone, there is no bringing them back. Just as we can't easily make more land, we can't easily make more soil. Soil is an undervalued, underdebated but magical and precious resource.

The other major concern is about availability of fresh water. It's a cliché, but just because we are on a watery planet, it doesn't mean that there is unlimited water. Water probably formed from condensed gases as the early planet earth cooled, and some may have come from crashed comets and asteroids. Precious fresh water is a tiny proportion of that supply, around 3% compared to what is in the oceans, and most of that fresh water is underground or locked up in ice caps and glaciers with less than half of 1% available on the surface. Although water cannot be 'destroyed' as such, it can be moved by human activity to places where it becomes inaccessible, and it can be polluted beyond use. Removing salt from seawater is possible, but at present hugely energy-hungry and expensive, so only a good prospect if it can be achieved with renewable energy such as solar power.

So those are the ultimate boundaries for the human race – set by availability of land, fertile soil and water, plus the extent to which we want to coexist with other species as opposed to driving them to extinction by hunting them or removing the places where they live. At the same time, we need to be concerned about not just the absolute availability of those resources, but also their health, which we affect with wastes from our activities. Pollution happens in lots of different ways and in different places, but some is global in reach and effects. The most prominent issues are climate change from the build-up of CO_2 emissions, the effects of ozone-depleting chemicals, the accumulation of toxins in the environment and in humans, and the loading of the environment with excess 'nutrients' from overuse of fertilisers.

How do these problems link to the stuff we buy? In

many and varied ways, direct and indirect. Here's a glimpse of what we need for stuff, and what's at stake when we take it.

Stuff needs raw materials

The first thing to consider is that more humans need more food, so more land has to be converted to agriculture to feed them. In a globalised economy, this land needn't be anywhere near the people wanting the food – staple foods such as wheat, rice and corn can be grown in many different places and traded on world markets (although the further food is transported, the more vulnerable it is to losses and waste from inadequate storage). And we use agriculture not just for food, but for other raw materials such as cotton and crops that are turned into fuel. As land is cleared for agriculture, other kinds of resources can be taken – timber, metals, minerals, fossil fuels. Sometimes it is the other way round, and building roads to access minerals or undertake logging leads to the land being cleared for agriculture. Either way, it is a case of man vs wildlife.

'Renewable' is a natural quality of animals and plants, not a statement of how we presently use them. To be able to renew themselves, they have to be left sufficient space and time. Forests are a theoretically abundant, renewable resource, but once virgin forest is cut down, it might take centuries to repair itself to a state where it can once again support a diversity of plants and animals. Forests are crucial because they help to regulate the climate. They keep water in circulation because as the rain falls on their (usually quite thin) soils, the plants suck it up, and let some of it evaporate back into the atmosphere, completing the cycle. If forest is cleared, this cycle is broken. It can cause changes to rainfall patterns hundreds of miles away, and allow the rain that does still fall to wash away the bare soils, causing mudslides when it rains or dust storms when the soil dries out. In Haiti, the almost complete deforestation of the country has made it vulnerable to frequent landslides, made even worse by the 2010 earthquake. The suffocating dust cloud that descended on Beijing in March 2010 was blamed on land degradation including deforestation. Based on these already observed effects, it is likely that the deforestation of large areas will ultimately affect the climate more broadly, although how this might link to climate change caused by build-up of carbon dioxide is unclear. There is another link, however, because cutting down trees and burning them (or allowing them to rot) means adding the carbon that they have absorbed during their lives into the atmosphere as well – the longer they stay standing, the longer they act as a 'sink' for that carbon.

The proportion of original wetlands that has disappeared is even greater than forests, as they are drained for agriculture and settlements. In the east of England, water is still pumped continuously from areas in Lincolnshire and around the Humber estuary, and agriculture is practised where there used to be rich wildlife, at considerable cost in terms of energy. It seems that fertile land is too valuable to allow it to be just wet.

Last chance to see . . .

Along with the wild places that are sacrificed in our quest for raw materials, we lose their inhabitants. Extinction is a realm that seems remote and generalised without some good pictures. One of the first conferences I helped to organise was about negotiating the UN Convention on Biological Diversity – all very dry and abstract until someone showed slides of the plants and animals that have disappeared in comparatively recent times. For the things that were driven to extinction prior to photography, and if no one thought to paint or draw them, the pictures have to be of the written variety. Bill Bryson, in *A Short History of Nearly Everything*, has some good ones, noting that 'it is a fact that over the last fifty thousand years or so, wherever we have gone animals have tended to vanish, often in astonishingly large numbers'. He recounts that early humans, once equipped with spears, managed to eradicate many types of extraordinary, never-seen-the-like-of-again large animals, including giant sloths, the glyptodon (a tortoise the size of a car) and, of course, mammoths. It didn't stop there. The 'natural' rate of extinction has been one species lost every four years on average; human-caused extinction may now be running at 1,000–10,000 times that level, which between now and 2050 could mean one species every twenty minutes.

Some of this extermination hasn't even been in the name of food or space to live, but a product of sheer carelessness. Bryson relates the story of the only example of a flightless, perching bird – a type of wren. In 1894, a lighthousekeeper, recently arrived on an isolated island off New Zealand, was brought a dead one by his cat, and sent it to the museum in Wellington. The curator got very excited, but by the time he arrived to investigate, the cat had killed them all. As Bryson observes, 'if you were designing an organism to look after life in our lonely cosmos, to monitor where it is going and keep a record of where it has been, you wouldn't choose human beings for the job'.

And so it continues today. As wild areas shrink – forests cleared, wetlands drained and coasts built on – there is less and less space for other species. There is a poignant account of some of those hanging on by their toenails in Mark Carwardine and Douglas Adams's *Last Chance to See*, updated twenty years later by Carwardine and Adams's friend Stephen Fry. The hard-won photographs of some of the most threatened animals on earth, including the African white rhino, the Amazonian manatee and the

Malagasy aye-aye (a type of lemur), together with the accounts of their imminent demise, are testament to our lack of progress in halting habitat destruction in the twenty years since Carwardine's first trip. In fact, it has got worse.

It is easy to forget that extinctions are about plants as well as animals, and about unremarkable animals as well as photogenic ones (the latter are known in the trade as 'charismatic megafauna'). No one ever got their cheque-book out for a worm. Yet the majority of species extinctions are of things we have never heard of, and may have disappeared before even being known to science. There are all sorts of reasons why this is a bad thing, including loss of genetic material that could provide medicines, new crop varieties and other economic uses. Our understanding of ecosystems is still rudimentary, so removing even apparently insignificant plants and insects could be like playing that game where you have a stack of wooden blocks, and have to remove random ones up and down the pile while trying to keep the whole edifice intact. You can get away with removing quite a few, but you never quite know what will cause it to fall over. The UN Environment Programme has a neat diagram of all the things that 'biological diversity' supplies to humans, and they are many. But most of all, it seems like an indictment of the human race that we can't be clever enough to understand and find ways to coexist with other species.

Stuff needs good soils

We clear wild spaces for agriculture to provide not just food but other raw materials such as textile fibres and, increasingly, biofuels. We will never win the battle to provide resources for a growing population unless we look after our soils. Soils are very old. They are the product of tens of thousands of years of rain and ice working to break up rocks into fine particles, rivers depositing rich silts, and plants and animals laying down their lives to replenish the 'humus' content in the soil. That stuff we treat as mere 'mud' is as venerable as the most majestic of mountains. Treated well, soil provides the life-sustaining growing medium for crops as well as pasture for livestock. Soil is where the planet's nutrient cycles do much of their work, using microscopic organisms to put carbon and nitrogen where they are needed to sustain life. But if overcropped, over-irrigated, overgrazed and generally overused, soil can become poisoned, useless and in the worst cases simply blow away on the wind. The US Depression 'dust-bowl' years were the consequence of bad farming allowing overworked soils to sit naked of plant cover and dry out.

Worldwide, loss of soils is a huge and growing problem. Soil erosion is lowering the productivity of as much as a third of the world's cropland. In some countries, such as Lesotho and Mongolia, erosion has reduced grain production by half or more over the last three decades. Soils are also damaged beyond use by over-irrigation in dry areas, the unaccustomed water allowing salt from the layers below to migrate upwards and poison the soil, eventually rendering it lifeless. Salt contamination has degraded large areas of Australia in the course of growing cotton and other thirsty crops, demonstrating how easy it is to mistreat soil, and how we do it at our long-term peril. As with many other environmental effects, the lack of care of the land is likely to interact with other problems. The seminal UNEP Environmental Food Crisis report of 2009 warns that unless we recognise and act on what we are doing, we will see global agricultural productivity drop by a quarter by 2100 as a result of land degradation, climate change, freshwater scarcity and an increase in invasive pest species, the last partly spurred by climate change making some areas warmer. Soils, in common with the other fundamental resources discussed here, both affect and are affected by other pressures, but the results may not be noticed until it is too late.

Stuff needs water

Which takes us on to water. In November 2009, England experienced unprecedented rainfall – in one area of Cumbria, a month's worth of rain fell over forty-eight hours, leading to extensive flooding and two deaths. By June 2010, after a hot spell and very low spring rainfall, the water company in Cumbria was talking about the possibility of hosepipe bans because of the high dependence on surface (reservoir) storage of water, which tends to evaporate. Countrywide, we know that three dry summers and winters in a row in England can lead to acute water shortage because of our limited ability to redistribute water from wet to dry areas.

The longer-term trend is not encouraging. In the UK, the increasing use of water in agriculture, irrigating crops that used to be rain-fed, will over the next twenty to thirty years lead to increased and severe pressure on water supplies. This pattern is repeated around the world – some places don't have enough water to keep people and industry happy; others have more than they can handle. According to the UK Met Office, extreme weather events, made more likely by global warming, will make this worse at both ends – drought and flood in turn. Unfortunately, no one can make any sensible predictions about what climate change will mean for rainfall patterns on a global scale. This puts huge pressure on developing better management.

Global consumption of water has been rising much more rapidly than the rate of population growth, meaning that average consumption per person is rising. But water is not a global resource, it is a regional one. Water is available within specific 'watersheds' (the areas that catch rain and channel it into rivers), so the limits on water use differ according to the location. In some places the limits are seasonal, depending on being able to store water through dry spells. In areas that pump water up from underground (mining water, in effect), the limits depend on how fast these ancient subterranean reservoirs are refilled. In some places it may be quickly from rain or melting snows, but in others it may take place over aeons, meaning that these are in effect finite water resources. Those delicious Egyptian new potatoes that appear early in the spring, for instance, are grown in arid soil with 'fossil' water pumped up from deep underground.

So there is not one world water crisis, but several. In her *No-Nonsense Guide to Water*, Maggie Black points out that even if on a global scale we should be able to manage with the amount of fresh water available, on a regional and local level, access to water and its rate of use are very unequal. Some of the poorest people in the world are in the driest areas, so if water consumption per person is rising around the world (including in the UK), we have to judge whether that is a good thing or not – in some places it needs to rise to meet basic human needs such as safe drinking water and sanitation, in others it is clear that rates of use are excessive.

Agriculture and industry consume far more water than people directly: globally, for every litre consumed in the home, two are used in industry and over eight in agriculture. As WWF points out, 'while average household water use in the UK is around 150 litres per person per day, our consumption of agricultural produce from other countries means that each of us effectively soaks up a staggering 4,645 litres of the world's water every day. Most of this is in the form of "virtual water", i.e. water that has been used to grow the crops that make the food we eat, the beverages we drink and the clothes we wear.' Construction materials also take large quantities of water to produce: one study estimated that a kilo of concrete has around two litres of 'embedded' water, for a kilo of timber it is around twenty litres; steel, 40, and plastic a massive 185 litres.

This rising consumption has been addressed largely from the wrong end. Most of the investments in water management have been to harness groundwater and surface river water, for instance pumping systems for aquifers and large dams. This has its own considerable impact in terms of energy and disruption. Very little is invested in conserving rainfall and achieving water storage in the soil (where, globally, plants get 70% of their water). Simple changes to soil management (as practised in places like West Africa and Israel) could make all the difference to the impact of the 100 million litres of water per second used in global agriculture.

The top water-using manufacturing industries apart from food and drink include mining, paper and pulp production, and computing. For mining it is for processing the ores, and for paper and pulp dissolving the tree fibre into the mash that becomes paper. By why computing? Microprocessor production needs ultra-clean conditions for the components to be free of contamination so that they work properly. This means repeated washing of the components, often with the addition of solvents to remove any grease, so a lot of waste water contaminated with chemicals is produced. Good practice dictates that this is captured, cleaned and recirculated, but as we've already seen, good practice doesn't happen everywhere.

Fred Pearce has chronicled some of the extraordinary levels of use and misuse of water around the world. There are the lawn sprinklers and swimming pools of Phoenix, Arizona, draining the life out of the once-mighty Colorado River to make a rich settlement cool and green in the middle of what should be desert. There is the grossly wasteful use of the waters of the Murray–Darling river system in Australia to irrigate cotton and rice – an Olympic-sized swimming pool of water to produce one bale of cotton in a country suffering severe drought. Most shocking is the deliberate emptying of the Aral Sea in Kazakhstan and Uzbekistan, causing environmental and human tragedy of epic proportions. The sea has shrunk to a tenth of its original size after Soviet-era planners diverted the rivers that fed it to irrigate cotton in near-desert areas. As a consequence, the salt content of the remaining water has soared, killing all the fish that once formed the livelihood of the people living around the sea. Much of the drinking water is contaminated by salt, and together with the dust storms of salt and pesticide residue that blow across the arid area, it causes anaemia, lung disease, cancers and birth defects. The climate has changed – without the moderating influence of the water, summers are shorter and hotter and winters longer and colder. As Pearce observes, it is hard to imagine worse mismanagement of water, but even harder to accept that the countries that inherited the catastrophe when the Soviet Union collapsed don't have any plan for repairing the damage.

Across the world, letting rivers run dry has irreversible consequences; it is not just a question of weathering a few dry years and hoping that the rains, and therefore river flows, will pick up again. Similarly, the overabstraction of fresh water from aquifers near the sea leads to 'saline intrusion' as seawater seeps into the aquifer and replaces or permanently contaminates the remaining fresh water. Throughout history, failing to store enough water against drought conditions, or allowing supplies to become contaminated, has been a factor in the collapse of civilisations. Not a happy thought, especially as we take our flowing taps so much for granted.

Stuff needs energy, and energy means climate change

Everything we have takes energy to harvest or to make or to run. This is the case whether we are talking about growing cotton and making it into a T-shirt, mining metals and assembling them into a mobile phone, or manufacturing plastic for the bag to put it in when we take it out of the heated or air-conditioned shop. On a global scale, most of that energy comes from fossil fuels, the remains of long-dead plants and animals. They contain carbon and they burn well, but they have an unfortunate side effect.

Some of that carbon (as carbon dioxide) ends up in the atmosphere where it adds to the earth's existing cosy blanket of gases, and it is gradually increasing global temperatures. Despite attempts in some quarters to wind up public scepticism, very few scientists now question that the planet is warming up, and that the warming is a consequence of burning fossil fuels. Recent years have seen enough warming to make the effect pretty much indisputable, and sea levels are already beginning to rise as ice melts at the poles. Clever people are working around the clock to attempt to model the effects of the warming, but climate and weather still defy our understanding to a worrying extent, so the models can never be taken as gospel. The future is uncertain, and it may not be good. But politicians worldwide are now acutely aware that climate change has potentially serious, but as yet not fully understood, implications for human activity – a situation that few politicians are well equipped to handle.

In March 2010, the Department of Health for England issued a statement on the health implications of climate change. These included an increase in heat-related deaths (over 35,000 excess deaths were reported from twelve European countries in the 2003 heatwave). It also said that 'increased coastal and river flooding will have major impacts on health – as well as threats to physical health, flood victims can also experience significant mental health problems as a result of personal and economic loss and stress'. At the same time, 'Episodic higher concentrations of ground-level ozone, caused by air pollution and often associated with heatwaves, could lead to an increase in respiratory problems and incidence of allergies.' Further, 'The impact of climate change on health will be felt around the world. The densely populated coastal areas of the world are at risk from rising sea-levels, whilst malnutrition in arid areas may be exacerbated by drought.' This hard-hitting statement was important for three reasons. First, it came not long after the media storm surrounding a number of climate scientists who had been accused of manipulating data to make the threat look larger than it is, and so represented a national government making an authoritative statement on the strength of climate science. Second, it came from the Health Department rather than the Environment Department, the former being a body with greater political clout. Third, it brought climate change up close and personal. I could almost feel myself choking on the heatwave-exacerbated traffic fumes as I read it.

Is there anything to be done? The international leaders assembling at the December 2009 international climate negotiations in Copenhagen were urged to make every effort to hold the temperature rise to two degrees Celsius above pre-industrial levels, mainly by moving away from use of fossil fuels. Beyond this, dangerous climate change is considered inevitable. To hold the temperature rise to two degrees, the scientists briefing the leaders advised that CO_2 emissions should be reduced by 60–80% immediately. That, of course, is impossible without ceasing just about all industrial activity and quite a lot of agriculture overnight. So the scientists advised trying to ensure that emissions should peak 'in the near future' and then come down. Although the UK has a target to reduce emissions by 80% by 2050, no global political agreement came out of Copenhagen to say when the turnaround is meant to begin, and there is no global deployment of alternative sources of energy sufficient to make the switch in the 'near future'. Economist Nicholas

Stern's highly influential 2006 report made a stab at showing the effects of temperature rises in one-degree increments and even at two degrees Celsius it is not looking good – increases in the intensity of storms, droughts, forest fires and floods, onset of the melting of the Greenland ice sheet and falling crop yields. But this is a best guess – we are still in very uncertain territory.

Stuff means pollution

Visible pollution used to be what dominated environmental discourse in the UK. Air, water and soil pollution have been consequences of industrial activity from earliest times, but accelerated during the Western industrial revolution, so that dirty factories, belching chimneys and soot dominate our images of the nineteenth century. In the UK, most of Europe and the US, pollution of the kind that led to blackened cities, acidified lakes, and rivers which ran different colours on different days depending on which dye was being manufactured, are largely (although not completely) a thing of the past. The same cannot be said for the countries that have taken on the mantle of being the world's factory, such as China and India, where a lot of our stuff is made. Here, in the absence of widespread or biting regulation, unbreathable air and contaminated water and soil are more common.

Some effects of pollution are, thankfully, reversible. In 1961, the Thames Survey Commission Board found that the stretch of the Thames running through London supported just four species of fish, having been used as a sewer for human and industrial wastes since the city was established. Today that number is 150plus, thanks largely to European legislation that has imposed limits on sewage discharge and industrial pollutants, and required the spending of millions of pounds on better management. That, however, doesn't stop the occasional large 'fish kill' when it rains hard and London's ageing sewer system overflows and starves the river of oxygen – 39 million tonnes of sewage still goes into the Thames every year. But we are in a good place compared with contemporary India and the once mighty River Ganges, for instance, where increasing amounts of sewage and industrial waste are rendering what used to be sacred waters desperately unsafe.

London also used to be famous for its black smogs, as much a product of household coal-burning as industrial activity. Edward I had to impose the first ban on coal fires in London in 1306, although clearly that didn't work for long. Towns up and down the UK were unbearably sooty and smelly up until the Clean Air Act of 1956. That is unimaginable today – all we have to remind us of a time when people lived in such unpleasant and unhealthy conditions are the dark deposits on buildings. Traffic fumes may be trying hard to take over as the chief cause of atmospheric irritation, but have a long way to go to emulate the cities of post-war Britain and the limitations that the conditions placed on health and happiness. Yet in China at the time of the Olympics in 2008, the order had to go out to factories around Beijing to stop pumping smoke out of their chimneys so that the athletes and spectators had a chance of breathing clean air and seeing across the city. Some of those factories make our stuff.

Some pollutants are not so visible. The 'toxic chemicals' agenda is a relatively recent one, and centres on the ability of the chemicals industry to create substances that have not previously been part of nature's cycles. But although to some people that makes all synthetic chemicals automatically suspect, this is not always easy to show. For most potentially hazardous chemicals, whether for human or ecological health, and whether of natural or man-made origin, toxicologists say that it is the dose that makes the poison. For any chemical, there will be a level at which ill effects start to show, because even substances such as vitamins can be toxic if taken in very high doses. By testing, a 'safe' dose can be established, and a safe level of exposure.

What is difficult is dealing with chemicals that may only have very long-term effects. The properties that make synthetic chemicals so useful are also what make them potentially problematic – like plastics, they are often based on oil, and, also like plastics, they are 'stable', i.e. they don't easily degrade back into their constituent molecules. Plenty of these have not been suspected of any ill effects, but some are very worrying – they persist and accumulate in the environment, and those that are also soluble in fat can 'bioaccumulate', i.e. build up in the bodies of animals and humans. This could have long-term and unpredictable effects on both human health and wildlife. Some of the 'POPs' (persistant organic pollutants) are suspected of being neurotoxins, meaning that they attack the brain and so might affect the development of children. Some scientists believe that an increase in neurodevelopmental disorders, such as ADHD and autism, may be a result of widespread exposure to low concentrations of multiple chemicals. But the effects of what may be a combination of different substances interacting with each other are very hard to pin down. Some substances, like the chemicals used to make some plastics soft, and which are suspected of interfering with human hormonal systems, don't even follow the normal dose/response pattern, and apparently have more effect at lower doses. The worry here is long-term effects on levels of male fertility as these substances increasingly circulate in the environment.

Some toxic pollutants have 'natural' origins but are still highly dangerous. This applies to many metals, which exist in the earth's crust but are only circulating in the environment in such high quantities thanks to man's extraction and manufacturing activities. They include mercury, lead, arsenic and cadmium, all of which are potent neurotoxins. Some historians blame lead for playing a part in the fall of the Roman Empire – apart from having lead plumbing, richer Romans cooked extensively in lead pots, not knowing that dissolved lead in food would slowly poison them, causing mental deterioration. This could be why Nero fiddled as Rome burned, and Caligula gained his reputation for being somewhat strange. A more recent and verifiable disaster has been the long-term dumping of mercury-polluted waste from a chemical factory in Minimata Bay in Japan, which has resulted in some 3,000 people suffering deformity, disease and death over a period of five decades. Even metals that we need as 'trace' elements in our bodies, such as copper, manganese, zinc and iron, can be toxic in large quantities.

One of the worst examples of this kind of pollution that I have seen was not in the developing world, but in Butte in the US state of Montana. An old mining town, Butte is home to the Berkeley Pit, America's biggest 'superfund' site – in other words, the largest industrial site subject to a vigorous legal dispute as to who should pay to clean it up. Open-pit mining for minerals and metals including copper, manganese, lead, arsenic and zinc has left a hole more than a mile and a half across and 1,800 feet deep. The pit is continually filling with water heavily contaminated by the metals in the remaining rock, and is also highly acidic. It is now around half full, and the water has to be constantly pumped away, and probably will have to be forever. Over the years, the waste earth and rock from mining (spoil) had been deposited around the pit and then homes built on it, a practice only recently stopped. We were told that some people's gardens had soil with concentrations of arsenic some twenty times above safe levels. Not far away is an ageing dam on the Clark Fork River, with toxic sediments from the mining piled up in the reservoir behind the dam, which have sent arsenic into local drinking water. No one in Butte seemed prepared to spell out what this meant for the health of the inhabitants. I went there in 1993 – seventeen years on there is still no agreement about how best to treat the continual flow of polluted water into the pit in the long term; all that can be done is to attempt to prevent it spilling out into the river or into people's homes.

It's not just the primary, 'extractive' industries that are responsible for pollution. The US's powerhouse of microchip production, 'Silicon Valley', has more EPA toxic superfund sites than any other area of the US, and more than 150 sites where groundwater has been contaminated – many of these are related to the high-tech manufacturing discharging toxic chemicals and traces of metals.

All these examples illustrate an unfortunate short-term business philosophy that worries about stuff first and environmental consequences later, if at all. Many contemporary companies are now doing a great deal to clean up past mistakes and misdemeanours, either because of state compulsion or concern about reputation, but very few companies throughout history have started out with a business model that took adequate care of the environmental underpinnings of their wealth. The fact that the limits to their operations are now being felt in terms of severely compromised quality of life, in countries rich and poor, developed and less developed, should serve as a warning about the future.

Stuff has unexpected consequences . . .

When it was first reported that there might be a 'hole' in the ozone layer in 1985, many people were unaware that there was such a thing. The ozone layer is the part of the earth's atmosphere that absorbs incoming ultraviolet light, the part of sunshine that causes you to tan, and also to get skin cancer if overexposed. The idea that man-made chemicals called CFCs (those invisible substances that powered aerosols and assisted dry-cleaning and flame-proofing) might be responsible for eating away the ozone was initially greeted with extreme scepticism by those in the industry. However, with unusual speed the science was accepted as incontrovertible.

The story of the discovery of the ozone hole is an interesting one from the point of view of human fallibility sitting alongside advanced technological prowess. NASA, the organisation responsible for sending men to the moon, was monitoring the concentration of atmospheric gases with high-tech planes and computer-data handling. The technology picked up the dramatic fall in levels of ozone, but since the results were so far outside the expected readings, the computer program rejected them as unfeasible and therefore probably instrument error. Three scientists, Joe Farman, Brian Gardiner and Jonathan Shanklin, from the British Antarctic Survey, sitting in a makeshift lab in Halley Bay in Antarctica, and using an instrument dating back to the 1930s, made the same measurements but drew a different conclusion. NASA has tried hard to cover its embarrassment on this one, but nonetheless it has been chalked up as a triumph for plucky British science. The lesson? Don't assume you have the environment taped, it can spring unpleasant surprises.

In the case of the ozone layer, identification of the problem led, unusually, to a swift political response. Within two years of the problem being linked to CFCs, an international law, the Montreal Protocol, had been agreed to limit their use; later, the Vienna Convention implemented more strict measures; and finally, the Copenhagen Amendments (1992) banned the production of twenty-one ozone-depleting chemicals. Today, most propellants have designed out CFCs, and the aim is to have a complete global phase-out by 2030. There has been enough of a reduction for scientists to project that the ozone hole may be closed again by 2040. This is reversibility over a long timescale, but modelling suggests that an estimated quadrupling of skin cancers up to 2100 from the greater exposure to UV has been avoided. We have breached an environmental 'limit', but may have got away with it by taking timely action. So where are the 'limits' that we really need to worry about?

8

Where the limits lie

Out of all the problems rehearsed in the last chapter, what most people want to know is: where are the absolute limits? This, like many other things, is a matter of judgement rather than hard-and-fast science.

If we consider that we take resources to fuel the economy, *and* the global population is growing, *and* (this bit is critical) we are not content with getting to a certain level of human 'welfare' and stopping, but we assume that economic growth must be continuous (the current recession in the developed countries is regarded by many economists as a temporary and mildly inconvenient blip), it should be no surprise that the picture does not look good. What do we mean by 'economic growth'? It simply means that the monetary value of the goods and services produced by a country, measured per person per year, rises all the time. It has been assumed for centuries that this continual rise is desirable, because it enables people to do ever more things beyond mere subsistence – for the state to provide welfare systems and go to war, for instance, and for individuals to buy more stuff and save up for the future. To get to such a point is regarded as being a 'developed' nation – to still be struggling to provide the basic necessities of life is regarded as 'under developed'. As a consequence of this assumption of continual growth, any account of likely total global resource needs as we go forward into the twenty-first century has scary numbers for everything we use: minerals, timber, fuel, food, everything. That was what Hilary Benn meant when he talked about the three planets – if everyone lived as we do in the UK, we would be using three planets' worth of resources. Certainly, there is no projection that shows demand for any of those things going down.

UK think tank the new economics foundation published a report in 2010 titled bluntly 'Growth isn't Possible', which featured the metaphor of a giant hamster. Hamsters grow very rapidly when young, but, like most other organisms, they reach a certain point and stop growing. They then consume resources in a stable way, not in an ever-accelerating way – in the wild, their access to resources would be dictated by living within the boundaries of the ecological niche they inhabit. In the view of nef and many eminent ecologists and economists, we should view the economy in the same way – something that we allow to grow to a certain point, then declare 'enough' and maintain a 'dynamic equilibrium'. Otherwise, the continued, accelerating consumption will be our undoing – in all the ways that we've just identified.

This line of reasoning makes perfect sense, and yet the report hardly made a ripple from the point of view of the mainstream media. This may have been because the image of a giant exploding hamster engendered mirth rather than panic, or perhaps because the news at the time was dominated by the allegations that climate scientists had manipulated data and overstated their case. It may be simply that it is very hard to visualise the effects of this unacceptable 'growth' on the ground. Or perhaps there is a feeling that we have had these dire predictions before, and nothing bad has come to pass . . .

Perhaps the most famous text in the 'we have a problem' part of the literature is the 1972 report *Limits to Growth*. It was produced by a team at the Massachusetts Institute of Technology, and commissioned by the Club of Rome, an informal group of international businessmen, statesmen and scientists. Based on theories about how systems behave and newly developed computer models, the report presented a range of scenarios around the consequences of continued growth in world population and growth in the global economy.

These scenarios suggested that human 'welfare' – the central goal of economic growth – could be seriously undermined by having to deal with the resource erosion and pollution created by that growth. 'Undermined' could mean anything from catastrophic collapse to unavoidable, but preferably seamless, adjustment. The crucial thing to the authors of Limits to Growth was whether humanity recognised the signs of being on an unsustainable path, or managed to 'overshoot' to the extent that the reining back would inflict serious damage to quality of life.

Subsequent commentators have criticised the projections in Limits to Growth as overpessimistic, arguing that the assumptions were based on the unprecedented growth of post-war Western economies, that the world has slowed down a bit since then, and there is less cause to worry about 'running out' of resources. On this last they may be broadly right.

What will we run out of?

It seems obvious that one day we will exhaust non-renewable resources. Metals, for instance, are finite – they are the gift of a turbulent process of planetary formation, a set of circumstances we can't easily recreate. But what is termed a 'mineral reserve' is a shifting quantity. For any given resource in the earth's crust there are three measurements. The first is what is actually there, which we can make estimates of, but don't necessarily know the full extent of. The second, which may be a lot smaller, is what we can readily access, i.e. is not too far under the sea, ice caps or built-up areas, or in politically inaccessible areas. The third, smaller again, is what is 'economic' to exploit – meaning that there is enough money to be made from getting the stuff out of the ground to make it worthwhile. The third measure, of course, is a moveable feast – if resources become scarce, the price goes up, making previously 'uneconomic' sources worth exploiting, and making it worthwhile to go looking for new sources. More stuff comes on stream, scarcity eases and the price (at least temporarily) goes down. Problem solved, for the moment.

For this reason, the 'reserve' of copper was estimated as thirty years in 1930 and ... thirty years in 2000. But although predictions of absolute 'scarcity' never seem to come to pass, there is a knock-on effect in terms of environments disturbed and waste generated. The concentration of copper in copper ore has decreased from around 12% in 1750 to less than 1% now, meaning that for every tonne of copper produced, the 'ecological rucksack' of wasted rubble and other materials is 500 tonnes. For platinum, it is now a staggering 500,000 tonnes, and it might even be economic in future to recover the platinum from street-sweepings, because the particles shed by catalytic converters mean that sweepings are just as concentrated a source as the ore (if you can lay hands on enough, of course).

The economics work in the same way for fossil fuels. We are already seeing the galvanising but damaging effects of much higher prices since 2007 in the exploitation of Canada's tar sands. These are vast areas of sticky, low-grade oil mixed with sand near to the surface in wilderness areas. This is difficult terrain to work and the product is expensive to produce, but the oil price has vindicated its exploitation. The trouble is, the price takes no account of the destruction of what were previously pristine landscapes. Even when environmental damage does appear to be limiting activity, there is always somewhere else to go: in 2010, at the same time as BP's share price dipped in response to its fatal rig explosion and massive oil spill in the Gulf of Mexico, rival oil company Premier saw a surge in response to a new oil find in the North Sea. Overall, the market is not signalling that oil has had its day.

The politics of minerals are just as important as their actual quantities. The UK is home to very few of the 'strategic materials' – mainly metals – considered vital to economic life as we know it. Reserves of important metals often come from parts of the world not noted for their political stability. This is at least in part because rich natural resources can be both a blessing and a curse. Minerals provide much needed income for both countries and individuals, and they are a blessing if exploited within an economy founded on stable government and the rule of law. But if these conditions are absent, natural riches can become the focus of corruption, conflict and severe inequality. Hence the growing awareness of 'blood minerals' – gold, diamonds, the coltan used for mobile phones – being sourced from war-torn countries such as Rwanda, Liberia and the Democratic Republic of Congo, and produced by impoverished labour. These problems may mean that we cannot count on being able to access mineral resources in future – they may be denied to us by those wishing to hold other countries to ransom, or we may wish to stop supporting corrupt regimes by buying their wares.

So 'insecure' is a broader concept than just 'scarce' – it combines physical qualities of metals, such as how easily they can be substituted with something else, and what kind of environmental impacts they have, with social factors, such as political stability and monopoly supply, as well as measures of rarity. On these measures, one UK report puts gold at the top of the list of the most 'insecure' metals, because it is relatively rare in the earth's crust, but it is also in areas with high political instability, and high vulnerability to climate change, and has high environmental impact to extract. Gold is also increasingly in demand as a 'safe' means of investing money after the 2008 banking crisis.

The so-called 'rare-earth' metals provide a good illustration of all these considerations. These metals (there are some fifteen in the group) are not all that rare, rather they tend to exist in low concentrations making exploitation often uneconomic. New extraction techniques in the 1940s lowered their cost, prompting new uses so that now they are considered important in a variety of very specialist applications. Ironically, some of these are in technologies aimed at reducing environmental impact, such as the neodymium used in the magnets in some hybrid-engine cars and in wind turbines, or the terbium used in low-energy light bulbs. The Chinese have increased their production of the rare earths so much over the last twenty years that prices have gone down and Western manufacturers have become dependent on them – now 95% of supplies are controlled by China, although this has been at considerable environmental cost. But the Chinese want these metals for their own technological revolution; they have gradually reduced export quotas and as of 2010 were considering stopping export of some metals altogether. This has sent other countries scurrying around looking for alternative sources, and financial pundits have recommended investing in firms in the US, Greenland and elsewhere as a route to future riches. In a bit of lateral thinking, researchers at Leeds University are purifying rare earths from low-grade titanium ore, the mineral that makes paint and suncream white, and which fortunately is much more abundant. Where there is a will (and shortage) there is a way.

So valuable stuff is still there, it is just getting harder to extract, and the issue is the environmental destruction, as well as the human cost, involved in getting it out. This is a major concern since mineral demand is the fastest-growing of all the materials, driven by global economic growth. It is a myth that the banking crisis put a brake on consumption, or at least anything more than a very temporary one. Global average growth for 2010 is projected as somewhere between 2 and 3%, and the dip into 'contraction' of economies worldwide (i.e. making less money from goods and services than they did before) lasted little more than a year. Our utilisation of minerals has improved considerably, getting more value from less resource, but still the absolute amounts brought into the economy continue to rise. This should put the emphasis on using the minerals already in circulation by recycling more. The upside of metals and other minerals is they may be finite, but they don't disappear: everything we've ever mined is out there somewhere, embedded in buildings, electricity cables, gadgets and a host of other products, as well as plenty in waste dumps.

Running down the capital

So, we're not necessarily running out of finite resources yet. But as Dennis Meadows and his co-authors argue in the thirty-year update to *Limits to Growth*, their report did not pretend to present one certain future, rather a range of possible futures, in which different levels of 'constraint' might come to bear on our activities. Three decades on from the original report, the authors stood by their original conclusion that the world is on an unsustainable path (even more so in 2002 than 1972), but also that we have choices about just how fast we hurtle down that path. Another update is due in 2012.

It is perhaps not surprising that, given the report's scary title, commentators concentrated more on disproving the scenarios than approving the authors' prescriptions for change. Meadows and his colleagues set out criteria for how to avoid overshoot, based on the thinking of Herman Daly, the key figure in the field of 'ecological economics' (this means economics that takes account of the economy's dependence on ecological systems, which most economics does not, which is why we have our linear economy). Nef's 'giant hamster' report also drew heavily on Herman Daly. The criteria are blissfully simple, and paraphrase roughly as:

1. We should not use 'renewable' resources (i.e. things that grow) any faster than they can grow back.
2. We should not use non-renewable resources (fossil fuels, minerals) any faster than we can develop alternative ways of doing the same job.
3. We should not put pollution into the environment any faster than the environment can deal with it.

This is clearly not how we are behaving right now. American environmentalist Lester Brown compares the modern global economy to one of those crooked financial schemes ('Ponzi schemes'), meeting current demands out of the earth's basic assets, rather than investing wisely and living from the interest. He writes: 'As recently as 1950 or so, the world economy was living more or less within its means, consuming only the sustainable yield, the interest of the natural systems that support it. But then as the economy doubled, and doubled again, and yet again, multiplying eightfold, it began to outrun sustainable yields and to consume the asset base itself. In a 2002 study published by the U.S. National Academy of Sciences, a team of scientists led by Mathis Wackernagel concluded that humanity's collective demands first surpassed the earth's regenerative capacity around 1980. As of 2009 global demands on natural systems exceed their sustainable yield capacity by nearly 30 percent.' So it is worse than just running up an overdraft – we are eroding our vital capital assets, the natural capital that underpins everything we do.

Even worse, we should be able to recognise that we are doing this, because previous civilisations have ignored these rules and paid dearly. Geographer Jared Diamond's book

Collapse charts how societies from Easter Island to the Mayans to the Greenlander Norse of medieval times undermined their very existence by failing to realise that the way they were using resources could not carry on forever. The Easter Islanders in the Pacific, the people who built the giant stone heads, apparently cut down every single tree on the island in a period between around AD 900 and 1600. This led to the loss of the wildlife that once provided food, caused catastrophic soil erosion that rendered agriculture almost impossible, deprived them of timber to build canoes for fishing, and meant an eventual descent into cannibalism and starvation. Exactly *why* they did this will never be known for sure, but Diamond presents plenty of evidence that the exercise of carving the stone heads from quarries and then hauling them miles to their sites and erecting them (some of them were 32 feet tall and weighed 75 tonnes) entailed massive use of resources. The people doing this intense physical work had to be fed, which meant clearing forest to make way for agriculture, which succeeded in supporting an increased population only until the resulting soil erosion undermined the effort. The making of ropes (from bark) and logs to haul the stones over required many trees. And why these gargantuan stone heads? Diamond thinks that they were a spectacular display of one-upmanship by rival clan chiefs, which may also explain why they got bigger and more elaborate as time went on. When the drain on resources meant that the island could no longer properly feed its population, the statues were toppled by warring clans inhabiting a desperate, disintegrating society. Similarly, the Mayans in Mexico, whose civilisation reached a peak of sophistication in the eighth century, had a fondness for lavish plaster decoration of buildings. This was made using copious quantities of charcoal that was made by felling surrounding trees, adding to the demands on the forest for fuel and construction. This led to soil erosion, depositing infertile soil from the hillsides on top of the fertile soils in the valley bottoms, reducing the productivity of agriculture and thus the ability to feed a population that had grown rapidly. Combined with drought and warfare, the effects of deforestation led to the eventual collapse of the Mayan civilisation.

These are simplified versions of complicated stories, but Diamond's central message is clear: collapse often followed surprisingly quickly after the societies had reached their apparent peak – in other words, the 'peak' represented not just the pinnacle of their achievements, but the product of unsustainable practices.

Ah, say the critics, but we are cleverer now. We have technology and we have accumulated knowledge. We know what we're doing. Diamond is unconvinced – for one thing, 'new technologies' are not always an unqualified boon. To be optimistic, we have to believe that all future technologies will do only good things, and stop adding to the problems. So there would have to be improved solar panels, but no further proliferation of energy-consuming gadgets, for instance. Most important, demand would have to remain static, while the supply was made more efficient. Doesn't our global economy offer some resilience? Actually, no, argues Diamond – it is just as likely that conflict rather than cooperation will ensue if resources come under pressure as beleaguered people move into others' territory. Plus the interdependencies caused by global trade and aid may lead to more widespread collapse – no country on earth is entirely self-sufficient in food and other stuff, so trouble in one place can mean trouble abroad. The 2008 banking crisis amply demonstrated this.

Can't we do things more efficiently? Unfortunately, greater 'efficiency' in the way we do things does not always relieve pressure on the environment. This is because of the annoying 'Jevons paradox' or 'rebound effect', which describes the tendency of efficiencies to lower the price of resources, thus leading to greater demand, and thus, paradoxically, to greater consumption. Jevons described this in the nineteenth century in relation to coal. New steam technology meant that the chemical energy of the coal could be more efficiently transformed into mechanical energy, so it became a more potent energy source, improving its cost-effectiveness, and demand soared. Indirectly, this 'rebound effect' has been observed in relation to householders. Buying a more energy-efficient fridge can be accompanied by putting the old one in the garage to store beer, on the argument that having one that is cheaper to run allows the continued use of the old one. This of course negates the overall energy savings. So greater 'efficiency' has to be coupled with other measures to limit resource use, a theme we will return to in Part 6.

Efficiencies can anyway be overwhelmed by sheer growth in demand, however that demand is stimulated. In fact, the world economy has become a lot more efficient in the way it uses energy and other resources; it is just that the savings don't register in the face of the huge rise in overall consumption. The ecological economist Arno Behrens observes that during the last century 'Materials use continued at a slower pace than the global economy grew, but faster than world population. As a consequence, material intensity (i.e. the amount of materials required per unit of GDP) declined, while material use per person doubled from 4.6 to 10.3 tonnes per year.' Roughly translated, this means that the economy got better at using stuff efficiently, but each of us used more and more (although not distributed equally across the globe, of course). As we noted at the beginning, with growing population *and* continued expectations of growing affluence, in both developed and less developed nations, resource use can only continue to rise. The hamster is still expanding.

Safe space

But when will it explode? As we saw above, the concept of a 'limit' to growth is a fairly subjective one, and embraces a world of different kinds of effects, so it is not easy to say. Some 'limits' are being experienced already, some may be a way off. In 2009, a group of scientists from Sweden, concerned with environmental resilience, decided to take a slightly different approach and work out 'an estimation of the safe space for human development'. They proposed nine 'planetary boundaries' which, if set with inbuilt precaution to take account of uncertainties, *ought* to guide future policies. In their words, they define the 'planetary playing field' for humanity if we want to be sure of avoiding major human-induced environmental change on a global scale. These nine boundaries are: climate change, acid oceans (a consequence of carbon emissions, where the increased carbon dioxide in the atmosphere ends up dissolved in seawater forming carbonic acid), ozone depletion, fresh water use, land use, biodiversity loss, chemical pollution and also two others that are more recently identified as problems – human interference with the nitrogen and phosphorus cycles (mainly through use of these elements in the artificial fertilising of agricultural land) and aerosol loading, the release of tiny particles from churning up dust and from burning fuels.

Of these nine boundaries, the authors conclude that mankind has already breached three. Two are overshot by a long way – the rate of biodiversity loss (extinctions) and the loading of the environment with excess nitrogen. The third is climate change, where we are edging into the zone where we risk it becoming an irreversible effect. Two others are at or near the defined limits – ozone depletion and ocean acidification. For two – aerosol loading and chemical pollution (considered on a global scale) – the science is too uncertain to define a boundary. That leaves just two with a bit of breathing space – fresh water use and land use changes. But reading the detail of these doesn't give much comfort, and broadly accords with what I described in chapter 7.

In a rather puzzling bit of editing, the New Scientist summed this up on its front cover as 'Earth's nine lives – why the planet is healthier than you think'. I have to say, that is not how it read to me. It perhaps demonstrates the dangers of drawing lines and then saying whether they have been crossed or not – human nature is to discount the ones that haven't. The authors gave heavy caveats to their calculations, proposing the boundaries as a method for thinking about the problems rather than as providing firm predictions, and they drew attention to yawning gaps in the data. Their central point was missed in much of the coverage – the subjects for the boundaries were chosen because of their potential to spring surprises. They might interact in unexpected ways. The dynamic, interconnected processes that give the earth's ecosystems their resilience might at the same time 'lull us into a false sense of security because incremental change can lead to the unexpected crossing of thresholds'. That suggests incorporating a good degree of precaution in deciding when to act. As renowned ecological economist Robert Costanza says, 'dealing with uncertainty about limits is a fundamental issue. If we are unsure about future limits the prudent course is to assume they exist. One does not run blindly through a dark landscape that may contain crevasses'. Or as one of the Limits to Growth modellers Jay Forrester put it more bluntly: 'Our only option is to choose our own limits, or let nature choose them for us.'

Back to stuff

We have seen that what we are running out of is not always visible in the shape of things we buy. What we are risking are the basic underpinnings of life on earth – forests, soils, water, together with clean air and uninterrupted global life-support processes like the nitrogen cycle, the ozone layer and a stable climate. We can't even really divide them into 'global' and 'local', given the complex interactions between them. Clearing forests may be felt to be a problem local to the nations that allow it to happen, but the knock-on effects in terms of climate change and disruption of rainfall may affect people a long way away. Conversely, global warming leading to climate change may exacerbate drought and local availability of water. Water pollution with persistent chemicals may be felt to be a local problem, but the chemicals move and accumulate in the environment such that people far away suffer ill health. Bits of plastic circulate in the Pacific thousands of miles from anyone – whose problem is that? Loss of species is an indictment of us all.

So how does all this link to stuff? In a great variety of ways. When we consume paper or timber, we don't always know if the forestry has been done sensitively or insensitively, legally or illegally. Anything we have that contains minerals, from the bricks in the house to the metals in the mobile phone, will have meant extraction activity somewhere in the world, and that activity could have been done relatively well, or could have wreaked havoc. The cotton in our T-shirts could have been grown organically, drastically reducing the energy, water and pesticides lavished on the conventional crop, but could be taking ten times as much land to get the same yield.

There is almost nothing we use that hasn't used energy and water in its manufacture. Glass has a cheap, abundant feedstock, but making it requires large amounts of energy that presently mainly comes from fossil sources, contributing to climate change. The same can be said of cement. Textiles need water to grow or process as well as energy. Plastic has a feedstock that will be available as long as oil, but here the impacts shift very visibly to the disposal end.

And woven through all of this are the disparities between impacts and opportunities in the rich and poor worlds. As nef has pointed out, 'by 'outsourcing' our manufacturing to less-developed countries, we are maintaining our environmental quality at the expense of theirs. And the outsourcing itself has additional impact – one of the key insights from a WRAP study on resource efficiency was that the carbon profile of stuff that is imported is often higher, as the emerging economies have had less time to achieve efficiencies. So it would have been better for us to make it here.

These issues are complex and connected, but we have to start somewhere. We have enough to go on when thinking about the future for the material world. Stuff needs to change. It needs to use less primary resource, particularly of trees and minerals. It needs to be much more sensible about using water in areas where water is under pressure. It needs to be made with less energy, and that energy must be non-carbon energy. It needs to be made using, and containing, fewer toxic substances, to reduce the legacy of harm for the future. It needs to be able to circulate in the economy much longer, saving the energy, water and materials needed for primary production, and avoiding stuff piling up uselessly as waste. We need to free land for food production, because that will be a key priority for the future, so the best non-food crop materials will actually be the wastes from food crops. Agriculture itself needs to reduce its dependence on fossil fuels and fertilisers, and look after soil and water. Few people could deny that all these things would bring benefits, but will they be enough to keep us within limits?

It is very hard to read the accounts of 'overshoot' and not conclude that some kind of cap on the use of physical resources is needed. A lot could be achieved with greater recycling and less waste, and all the measures mentioned above, but whether it will reduce overall demand enough to keep us within 'limits', wherever they are drawn, is still a moot point. We may need to decide how much stuff the planet can afford to give us, and what each of us is entitled to. When should the hamster stop growing? The UK government has set carbon budgets under the Climate Change Act to achieve a big reduction by 2050 – perhaps it is time to consider setting resource consumption budgets.

Of course, it is very hard to take a step back and ask of global economic activity: how should we organise this differently? It is much easier for each of us to ask ourselves: how should I behave differently? The latter approach seems to me predicated on the notion that the problems are largely down to us. They may be a consequence of our cumulative consumption, but how much of that can be laid at our door as individuals? The next two chapters give my answer.

Part 4
Us and Stuff

9

The curse of the giant pink pencil

I am in the office having a chat with Tracy. On her neatly-ordered desk is a giant pink pencil, some twelve inches long and an inch or so thick. It has an appropriately giant rubber fixed in the end, and the name of a conference company printed in black letters down its side. It has what appears to be real graphite as the lead, and a neat conical piece of rigid plastic to protect the tip. It must contain at least as much timber as thirty normal pencils. I pick it up in wonder. 'Isn't it great?' says Tracy. 'But surely you can't use it?' I respond. She looks crestfallen. 'I know, but I have a bit of a thing about giant stationery.' I open my mouth to harp on about the pointless use of resources, but decide not to spoil her day. But why did anyone think that was an appropriate marketing device to send to a green organisation? Or indeed any organisation? Presumably because they know that there are people like Tracy who appreciate giant stationery, coupled with an assumption (almost certainly erroneous) that the recipient will think of that company before all others when considering where to hold a conference. And also because, from an economic point of view, giant pink pencils don't cost that much to produce. A couple of months later I see it again and the rubber has parted company with the rest of the pencil, leaving Tracy with the dilemma of which of the many office recycling bins to put it in. Her pleasure in the consumption of giant stationery was not long-lasting.

We are all consumers. Save for a few tribes still living by hunting and gathering, or people who have opted out of modern life in favour of self-sufficiency, most of the humans alive on the planet today are consumers in the sense that we most commonly understand it, i.e. they buy things, rather than produce everything they own or use for themselves. Most of us would like to think that we're much more than consumers – parents, children, siblings, spouses, partners, participants, citizens. But even these roles are often defined by patterns of consumption – the acquisition and replacement of things.

Without consumption, the economic system as we know it could not function. The idea of capitalism is that there is private enterprise (as opposed to state control of the economy as in communist regimes) and people make things to sell to other people. People are paid to make those things, so that they can buy things made by other people. The essence of capitalism is that firms are allowed to make profits in the course of doing this, which they either reinvest in the firm, save for a rainy day, hand out to shareholders, or use to make their directors richer (the last one helps them to buy more things). In the developed economies of Europe, the US and Japan, large numbers of people have moved on from actually making things (manufacturing) in favour of jobs in the 'service industries' or 'public sector', the latter being activities the state can afford to provide (schools, health service, etc.) because of the taxes levied on the money earned in the course of selling things. But things still have to be made somewhere, and someone has to buy them. We cannot get away from the fact that the economy relies on real stuff – materials and the products that can be made from them. These are the 'feedstocks', the underpinnings of modern capitalism, just as they were of the ancient systems of bartering and trading, and their consumption is inescapable.

As I sketched in Chapter 3, 'mass' consumption as we

know it took off after the Second World War. The Victorians started it, of course, by ramping up the Industrial Revolution, enabling a lot more stuff to be made more quickly by fewer people. The British Empire helped the process along by providing access to resources in far-flung places, lots of which we took without asking (or indeed paying). Shopping started to be a recreation from late-Victorian times onwards, but the real escalation in the availability of products, and the income to pay for them, was a phenomenon of the late 1950s and early 1960s and has continued vigorously until now. All along the way there have been critiques of 'consumption', from political and social as well as environmental standpoints. It seems to me that in the first decade of the twenty-first century the voices of warning have increased, and are beginning to form part of mainstream thought.

Turbo-charged consumption

One of the first books I read on this theme was Vance Packard's *The Waste Makers*. It was written in 1960 but could have been written today. 'Prodigality is the spirit of the era. Historians, I suspect, may allude to this as the Throwaway Age.' He complains that the US, having constructed a huge military-industrial machine employing a large number of people, had to find a way to keep that edifice intact once the Second World War ended. The answer was to sell people things they didn't need. The effects of this, he argues, were to impoverish important aspects of people's lives, including their relationships and their spirituality. For this degradation of the lives of US citizens he blames the government and the advertising industry – anyone who is a fan of the TV series *Mad Men*, about the driven, shallow, womanising advertising executives of 1960s New York, can get a flavour of the kind of thing he had in mind. I searched *The Waste Makers* in vain for any mention of waste in the physical sense. Packard doesn't seem to have been worried about where those things end up – his lament is about what it does to our souls to buy things we don't need.

Fifty years on and many are complaining of the same malaise. Both before and after the credit crunch there have been plenty of authors willing to take on the "affluent society" as something corrosive rather than enriching. Each attempts an understanding of *why* we go down this road of evermore stuff as well as considering the consequences. My fantasy dinner party comprising twenty-first-century consumption critics would include writer and philosopher Alain de Botton (*Status Anxiety*), think-tank leader Neal Lawson (*All Consuming*), journalists John Naish (*Enough*) and Richard Girling (*Rubbish* and *Greed*) and psychologist Oliver James (*Affluenza*). For extra insight, I'd include the neuroscientist turned economist, Pete Lunn (*Basic Instincts*).

Despite being concerned about consumption in various ways, I'd have to hope that none of them would turn down the chance of a free meal in return for helping me to understand the root causes (Richard Girling gives a particularly graphic account of eating a three-course meal that he didn't really need). And I probably couldn't get away with an Evie and Ed-style 'sneal' either – it would have to follow current culinary norms. My instinct is that Neal Lawson would be first off the blocks in the discussion, before even the starter had been served, because his book reads as if he sat down and started typing and didn't stop until he'd got it all off his chest. He believes we have been sold not just consumerism, but what he calls turbo-consumerism, a 'hamster wheel' that we're terrified of falling off. We have a marketplace that operates twenty-four hours a day, in shops and online, and invades places it has no business to be, including childhood and schools (he is particularly cross about the early sexualisation of children through the marketing of 'mini-adult' products).

Worse, we are willing hamsters. We have allowed ourselves to be beguiled by marketeers to spend time shopping which would have been better spent with our families or on community activities, we have allowed shopping to take the place of religion, and we have amassed so many material goods that there is a booming industry in storage for the overflow. Even politics is becoming a marketplace, selling us personalities rather than principles. And as if this wasn't enough, he fumes, shopping is literally costing the earth – in climate change, deforestation and species extinction.

Alain de Botton, in his gently erudite way, would contribute that he shares the concern about the devaluing of our lives – his definition of 'status anxiety' is 'a worry, so pernicious as to be capable of ruining extended stretches of our lives, that we are in danger of failing to conform to the ideals of success laid down by our society'. John Naish might nod in agreement, and elaborate the 'pointless tread-mill' theme. He thinks we have had enough stuff, enough food, enough information, enough work, enough happiness (at least of the kind pedalled in self-help manuals). Certainly, in terms of both personal and planetary ecology, we have had enough economic growth. He quotes the economist John Maynard Keynes, who helped solve the unemployment problems of the 1930s, and who thought that by now (i.e. three generations on from his own) we would have solved the economic problems that dogged the pre-war world, that we would have all we needed materially and would be free to move on to higher things: 'we shall honour those who can teach us how to pluck the hour and the day virtuously and well'. At this point, Oliver James might interject that how we live now could not be further from that ideal, and that what we're talking about has the status of a disease, which he names 'affluenza'. It is spreading uncontrollably across the planet from the UK and US, and is responsible for rising rates of what he calls 'emotional distress' and others call 'mental illness'.

Lawson lays the blame for the dramatic acceleration of consumption and materialism of the last three decades firmly on the Thatcherite free-market consumer-capitalism ideals of the 1980s, and the encouragement of the 'me generation', compounded by the fact that governments and financial institutions have been prepared to let people get into unprecedented levels of debt in the name of fuelling growth. Culturally, we are given no alternative but to remain on the hamster wheel. Oliver James similarly blames a particular brand of 'selfish capitalism', akin to Thatcherism and Reaganism, characterised by corporate short-termism, privatisation, liberalised markets and the promotion of materialism. These forces have allowed the rich to get richer, and everyone else to try to emulate them. He defines 'materialism' as 'placing *too high* a value on money, possessions, appearances' and he contrasts this 'relative materialism' with 'survival materialism', which is OK because it 'entails the meeting of a fundamental human need', but as he admits, the precise point at which materialism becomes relative 'is not clear cut'.

Alain de Botton goes a lot further back in history to examine a slow progression from medieval societies where everyone was allotted their place in life from birth and had little option but to put up with it, which at least meant a measure of certainty and security, to the rise of democracy, meritocracy and capitalism, where the idea is that if you are a failure it is no one's fault but your own. Thus greater equality of opportunity and higher standards of living are accompanied by continual anxieties about job security, acquisition and self-worth. He observes that 'a firm belief in the necessary misery of life was for centuries one of mankind's most important assets'.

But these forces of liberalisation had to have something to work with, which in Lawson's view is the human tendency to be tribal, wanting to 'belong'. Only this explains the tendency to consume for show rather than for need, as well as the way that brands exercise such power over the spending habits of individuals. Pete Lunn might chip in here to agree, because as he points out in his book *Basic Instincts* about behavioural economics (i.e. the economics of the real world, as opposed to theory), if people were really the 'rational, selfish individuals' that classical economics holds them to be, they would be far too savvy to fall for marketing blandishments, or indulge in conspicuous consumption. It is only because people are programmed by evolution to stay with the group, and at the same time want to demonstrate their status within that group, that they believe that a particular name attached to a product can encapsulate something important about who they are. Naish also attributes the drive to consume to the 'want circuits' in our 'instinctive Stone-Age brains', which are overexcited by all the goodies on offer, and are not always over-ruled by the 'higher' bits of our brains that are capable of more considered thought.

Richard Girling might pause before the next forkful of his main course to lay the blame for greed in *all* its manifestations (food, sex, material goods, land, power) firmly on our genes. We are wired to consume, more and more if we are allowed to, until something gives – 'given the chance, we take all we can get rather than what we need.' The 'it ain't our fault, gov, we're made this way' theme runs through many of these texts like the red stripe through toothpaste, while we try to cling to the hope that there is part of us that might transcend these primitive urges. Oliver James, on the other hand, sees this peddling of evolutionary psychology as a cop-out, a way of making us feel comfortable with what is essentially a political phenomenon. We have been urged to consume because it suited a post-war model of prosperity.

Over pudding, Lawson, Naish, James and de Botton would all shake their heads and agree that far from making us happy over the long term, after the buzz of buying is over, we are increasingly miserable. We don't seem to know what's good for us. Girling is the only one who seems resigned to greed: 'we have the intelligence to override our genetic impulses, but we cannot delete them like unwanted software. We can modulate our greed but we can't escape its gravitational pull. Snouts were made for troughs.' Naish, though, optimistically recommends meditation as a way to appreciate what we have, live in the moment rather than continually chasing some supposedly reward-laden future, and get off the consumer treadmill. He thinks that if we can only cultivate our 'higher brains' we might pull back from the idea of continual growth and be satisfied. Ironically, however, we would also have to involve the more primitive brain and appeal to herd instinct to get everyone to do it together. Lawson acknowledges that we can't just stop shopping completely, but he wants to find a better balance between the roles of consumer and citizen. He wants consumer boycotts, voluntary 'downshifting' of lifestyles, and a legal framework for a 'post-consumer' society, which would include curbs on advertising, taxation of luxury goods and waste, and even rationing. Overarching this, he wants progress to be measured not by economic growth, but by a 'well-being' index. James wants much greater support for parents so that they can stay at home with children during their early years, which is when he feels parents who are off working and chasing material-istic goals do the most damage.

By the time we'd all finished eating we'd be agreeing that humans must shed their adherence to stuff as a way of making themselves feel better and measuring their prowess. But what's this? Anthropologist Daniel Miller has turned up for coffee, and is about to turn the evening's discussion on its head. Miller is one of anthropology's leading lights on 'material culture' – what things mean to us as social animals. His books *The Comfort of Things* and *Stuff* put forward an alternative view of our rela-tionship with the material world, as one that can be enriching as well as impoverishing. He argues that having stuff, or having more than other people, does not make us bad people, in particular not necessarily bad at relationships. In fact, in some of his examples, having no stuff is a sign of a severely diminished life, in physical, emotional and spiritual terms. Miller explicitly puts up this thesis as a challenge to environmentalists' responses to consumption, arguing that they are based on too narrow an understanding of the role that stuff plays in people's lives. I can identify with this. One of the most distressing aspects of my mother eventually agreeing to move into residential care (four years on from the maggot-infested wheelie bins) was that she wanted nothing from her home. It meant not just that she didn't need much, but that she had let go of many interests and attachments, as represented by the objects around her. Later, I had to clear her home (which had also been my home for a good part of my life) of fifty years' worth of belongings. It is impossible to do this without feeling that you are somehow wiping the person away.

To round the evening off, Miller has brought Tim Jackson, one of the leading thinkers and writers on what 'sustainable consumption' might mean. He explains that, yes, all these explanations are credible. The insights from evolutionary psychology do suggest that we are driven by sexual and social competition, with consumption of stuff as a means of display and indicating status. Even if we are aware of this, and unwilling participants in the game, we are 'locked' in by political systems and social norms that give us very little individual control. And yes, as Miller and Lunn argue, the importance of 'stuff' in forging our identities and signalling our membership of groups is so strong that simply telling people to consume less doesn't work. Unfortunately, on top of all that, plenty of research over many decades confirms that in the West, since the war, we are richer (as measured by GDP) but, overall, no happier. We bid each other farewell at the end of a long evening feeling a little lost in the magnitude of it all, but very well fed.

'Wants' and 'needs'

The next morning I am still trying to work it out. We have a number of explanations for 'consumption' and 'consumerism', and a choice of reasons for feeling that 'excessive' consumption is a bad thing, and we also have ways of telling ourselves that stuff is important. But to me the key question is: *who* is to say what is 'excessive' and how much we really 'need'?

No one dares to try and answer this question. It raises the spectre of conflict and totalitarianism, since it is in those circumstances that governments feel they have a mandate to tell people how much they can have. Constraining people's freedom to consume is something that we have rarely accepted in the capitalist world; only in times of war when there is a common enemy and unprecedented collective effort does individual greed become seen as inappropriate.

My mother always contended that many people professed to be happier in these circumstances, herself included. They had a clarity of purpose and uncluttered lives, despite living with privation, fear and loss. People put up with rationing until nearly ten years after the war, but by the time the economy picked up and stuff was available again, including new kinds of stuff, they were clearly ready to embrace it. How long can people live straitened lives, if they have any choice in the matter?

I was deeply influenced as a teenager by reading Ursula Le Guin's *The Dispossessed*. The people in the story have been exiled from an earth-like planet because of their belief in a charismatic leader, and are forced to live on a barren moon with barely sufficient resources. They cope by creating a society where nothing is owned, even the children are looked after communally and talk of 'the' mother rather than 'my' mother, and any display of unnecessary stuff is labelled 'excremental'. In pursuit of scientific progress, one of their number is allowed to revisit the home planet, where he is shocked by the waste and greed. There is a lovely passage describing his incredulity as he watches a small present being wrapped in layers of packaging. At one point we learn what happened to the planet these people call Terra (Earth) – it is described as a ruin, a planet spoiled by the human species, where there are no more forests, the sky is grey and it is always hot. The population has been reduced from 9 billion to half a billion, and Terrans have only survived by imposing a totalitarian society. The book was written in 1974 so Ursula Le Guin was one of the first people to put the possibility of global warming in front of her readers, and I remember it sent a shiver down my spine at the time.

Despite the grim portents of the story, I enjoyed the book as I have enjoyed few others. It played to my forlorn hope that it is possible to be happy in a world of few things. The trouble is, I can't really put my hand on my heart and say I believe that.

Confession time

I can't believe it because I love shopping. I have trained myself to indulge in a particular kind of shopping – anything second-hand or sold for charity gets first call on my spending – but it's still shopping. I browse charity shops for diversion and recreation, and for the thrill of finding something I can just about justify having and then getting it for pennies. But that still means I buy things I want but don't really need. Periodically, I have a clear-out and give a lot of it back to the very charity shops it came from. My husband is the same, and I have a suspicion that a lot of my shopping is to keep him company and justify his habit. I can go on charity-shop expeditions on the slightest excuse – a gap in meetings, needing a present for a friend (although it is wise to keep the source a bit quiet) or a treat for the children. Pretty much my entire wardrobe is second-hand. And then there are books. I have a serious book habit, well catered for by second-hand bookshops, online marketplaces, car boot sales and friends. Every time I acquire a book I feel I have gained some important piece of insight or a new skill.

So speaking personally, I completely understand the urge to consume, and have no trouble seeing it as a creative activity rather than a hollow one. Shopping is fun. I don't relish stopping shopping, although I am happy to complement the process of consumption with the process of continual recycling. Perhaps consumerism is only bad for us if we allow the drive for continuous acquisition to make us miserable, and squeeze out other important aspects of our lives – family, friends, spirituality, smelling the roses.

But it is clear that we can only accept continual consumption if it can be somehow reconciled with the environmental limits already discussed. Keeping within the limits might give us a kind of 'consumerism' that is also saner and healthier, one based on a heightened awareness of the riches at our disposal and what we risk if we abuse them. How will it happen? Will it be the people labelled as 'green consumers' who will drive the economy in the right direction?

10

The myth of the green consumer

There are those who think that the key to a better world lies with us as consumers. One of the best books I've read about the relationship between the rich and the poor world is Giles Bolton's *Aid and Other Dirty Business*. Bolton has worked as a diplomat and with NGOs, and has spent time in Kenya and also Rwanda in the immediate aftermath of the genocide of the 1990s. He is clear that Africa has been disadvantaged not just by colonialism, but by recent aid policies that act more in the interests of the donors than the recipients (such as food aid that is effectively dumping the products of misguided agricultural subsidies for US farmers), as well as the terms of trade drawn up by rich countries for the benefit of rich countries. But he is also clear that when it comes to trade, we as consumers have a big role to play in forcing governments and business to pay attention and change what they do, and he can't understand why we don't use it more. 'Our failure to realise our clout in a consumer-fixated world, is, in truth, the most baffling aspect of modern life and at the root of our powerlessness'.

Jared Diamond likewise believes that a great deal of responsibility lies with the individual, as citizen and consumer. He chronicles the destructive practices of logging and mining companies, and concludes: 'In the long run, it is the public, either directly or through its politicians, that has the power to make destructive environmental policies unprofitable and illegal, and to make sustainable environmental policies profitable.'

But being a truly effective 'green' consumer relies on us *knowing* what's going on. And generally, given the complexity of the impacts of consumption, we don't. Being a 'green' consumer on presently available information is very much about making individual choices about specific products, usually against a background of knowing very little about the majority of products on offer, and certainly knowing only very general things about their cumulative effects. The link between 'consumption' and environmental problems and 'limits' is hardly ever clearly made.

It also relies on us *caring*. There was a time when it was predicted that green consumers would change the face of buying. Julia Hailes, who wrote the first *Green Consumer Guide* with John Elkington in 1988, declared: 'demand green products and you help develop new market opportunities for manufacturers and retailers – encouraging them to invest in new products and services specifically aimed at the Green Consumer'. Unfortunately, that will only work in a comprehensive way if all consumers *decide* to be green, seek out the relevant knowledge, and influence their friends and families. But not everyone demands green products. Those who do have become, in effect, a niche catered for by a few specialised products. Meanwhile, there is still plenty of investment in products going in completely the wrong direction.

Today, alongside Julia's work, there is a plethora of green consumer 'self-help' guides. If you have the time and inclination, you can find out where to buy ethical jewellery, how to plan a green funeral, and which companies are currently subject to ethical boycotts by one group or another. There are also numerous websites offering advice and goodies – the Internet has made green consuming something you can do in your organic cotton nightie. It's not that any of this advice is wrong, or not worth doing, but in the day-to-day world of retail, none of it feels mainstream, or, at the risk of offending dedicated green consumers, 'normal'. There is too much to read, too many alternatives to weigh up. There is too much to *decide*.

If I put my 'green consumer' hat on and take a trip to the supermarket, there are a few products that might leap out as meeting my criteria. They include organic food, fair trade goods, phosphate-free detergents (sometimes just labelled as 'green') and recycled toilet paper (or the ones labelled FSC which means it is from a certified 'sustainable' source of wood pulp). More recently, there are a few products labelled with their carbon footprint, including crisps and shampoo. If I am at the garden centre, I will be seeking out peat-free compost. Most of the rest of the stuff I need to buy each week comes without comment of a green kind from its makers. What am I to infer from that? Green credentials to be taken as read, or non-existent? Certainly I can opt for largely organic food if I can afford it, but entire shelves in a modern supermarket (pharmaceuticals, clothes, household bits, CDs, drinks, stationery, beach BBQs) will give me no clues as to their origins and impacts, except, possibly, in relation to the packaging.

You will notice that for the items on my 'green' list, there is an alternative to the 'green' choice. That leads me to feel that being a 'green consumer' is something I can choose to demonstrate by buying particular products, but if I don't want to, that's fine. Of the people who join me at the checkout queue, how many have made the 'green' choice? Not all that many from what I can see.

I'm going to briefly consider four of those 'green consumer' options, and try to illustrate why it is difficult for consumers to influence them sufficiently.

Clean and green? Phosphate-free detergent

This is an area where the complexity of the debate makes it very difficult to boil down the choices into ones that the 'green' consumer can influence through buying specific products.

The first detergents (the word simply means something that cleans) were forms of soap made as much as 3,000 years ago. The Romans knew how to make soap out of animal fat stewed up with wood ash, a strong alkaline agent. The fat/alkali reaction gave the soap the magical properties of being able to combine with grease and dirt, but still be soluble in water, so that the dirt was separated off. Synthetic detergents were first developed during the First World War, and were increasingly used after the Second World War owing to shortages of animal fats. These new detergents were developed from coal tar, then petroleum, and then also vegetable oil to make the 'surfactants' (the bit that does the dirt separating), along with phosphate-based 'builders', which make the surfactants work more effectively and stop the removed dirt depositing back as a uniform grey layer on the washed fabric.

Phosphate is one of the mineral nutrients essential to all forms of life and is thus essential to us. Mineral sources of phosphate, in fertilisers and animal feeds, underpin modern agriculture and thus our diet. We then excrete most of that phosphate in our sewage, to which we add additional phosphate from our use of detergents as well as from industrial uses such as food additives and making toothpaste. The phosphate then moves from being a useful fertiliser and cleaning agent to being a potentially polluting chemical that standard sewage treatment works can only partially remove – to do so completely entails extra treatment. The phosphorus in sewage effluent, along with run-off of manure or fertiliser from agricultural land, can lead to overenrichment (an excess of nutrients) in rivers and lakes, by providing nutrition to algae and plants that would normally only get very limited sources. This can trigger uncontrolled growth of algae, and algal 'blooms' became a big problem in North American and European lakes in the 1960s and 70s, using up the oxygen in the water, killing other fish and plant life, and altering the types of plants and animals that can live in that environment.

The problem of excess nutrients is called 'eutrophication', and although it has never been as great or as visible in the UK, it does still affect our water quality. A sparkling upland lake may become cloudy and suffer from toxic or nuisance algal blooms. The same amount of phosphate in a flowing river might have little visible effect, although it may alter the make-up of the plant communities in the river. The RSPB is concerned about the knock-on effects on birds, which can suffer from the decreased diversity of aquatic life that can accompany eutrophication. But before everyone starts blaming bird declines on washing powder, most of the phosphate in rivers comes from human waste (sewage effluent) rather than detergents, and around a quarter comes from run-off from fertilisers and manures from agricultural land. The contribution of these different sources will vary depending on whether the water reaching the river is coming from an urban or an agricultural area (the former will contribute more from sewage, the latter more from fertiliser run-off). I did say it was complicated.

Phosphate in detergent does its job very well, but there are alternatives, and some 'green' brands that were also marketed as 'phosphate-free' came onto the market in the late 1980s. These found a small uptake among consumers, but never dominated the market. Meanwhile, many major brands formulated away from phosphates anyway, thinking that eventually they would be subject to regulation. Either mandatory or voluntary restrictions on phosphate in detergent have been introduced in several EU countries, particularly to limit phosphate in domestic laundry products (stuff for clothes washing, as distinct from dishwasher or washing-up products). The amount of phosphate in laundry detergents is now generally limited to a very low 0.5% (in the past it had been as high as 30%), and the government has recently consulted on making this a mandatory level in England and Wales from 2015. Already 80–90% of UK laundry detergent products are made without phosphate. For the future, the government will look at the potential to restrict phosphate in dishwasher detergents if it can be satisfied that the alternative chemicals are safe and effective. There have been reports that some of the substitute chemicals are themselves causing problems; this is being researched, but the concerns have not been sufficient to halt the shift to low phosphate for laundry products.

Phosphate is an issue where it is almost impossible for the consumer to work out exactly what is the 'right' thing. Detergents are by no means responsible for the whole problem; they only add to the load in those places where there is no removal (stripping) of phosphates from sewage treatment works (about 50% of works) or no mains sewerage (where properties are connected to septic tanks or package treatment plants). But out of these, since only about half the UK population discharges its sewage to waters that are sensitive to phosphate enrichment, it's only detergents from at most 20% of the UK population that might be contributing to a problem. The industry agrees that eutrophication needs to be addressed but thinks that phosphate removal is best done at the sewage works, and that the phosphate alternatives developed so far result in products which are less sustainable and have poorer environmental profiles. Another twist is that with biological phosphate-stripping technology, the phosphates from both sewage and detergents could be recovered and used as fertiliser or even made into new detergent. This is done in some places in continental Europe and North America, but so far in the UK has been considered uneconomic and the approach has been to use chemical phosphate stripping, which is considered more reliable and less energy-intensive. Although the sludge from chemical stripping can be put back on land in carefully moderated amounts (the UK uses 70% of its sludge in agriculture), the ability to recover additional phosphate through the better technology could become very important, since recently the issue of long-term availability of mined sources of the fertiliser has come to the fore.

Phosphate has to be mined from rock deposits and easily accessible supplies may be gone in fifty years' time. There are huge reserves in China, but in 2008 the Chinese banned export of phosphate to protect supplies for their own agriculture. Although the detergent industry only accounts for around 10% of phosphate use, this has prompted some people to question the wisdom of using a resource that is also a valuable fertiliser simply to make our washing easier, and then failing to recover it (as well as the phosphate from sewage). On the other hand, with phosphate recovery at sewage works, all the phosphate from our food as well as detergents is potentially recyclable.

Where does any of this leave consumers in terms of judging the merits of 'phosphate-free'? If we're buying laundry products, the industry does seem to have responded to the combined signals of consumers and regulators and developed new products, though reluctantly. But dishwashing products are lagging behind – they now contribute slightly more phosphate than laundry products, and use of dishwashers is increasing. How are we to know about any of this and make the right decision? And is there a right decision, given that dealing with the phosphates problem spans water treatment policies and costs, agricultural policy, and ultimate resource availability? What is really needed here is a systematic approach to changing the ways we use and abuse nutrients. We will come back to some possible solutions in Part 5.

Down the pan

Sometimes the choices are simply too many. Into this category, I put toilet paper.

According to Rose George the author of *The Big Necessity*, the world divides, culturally, into wipers and washers. The Romans used washable sponges on sticks; the Chinese have long used paper like most of the West, but large parts of Asia have never done anything other than wash themselves with a pot of water.

Everything from dried leaves, through to moss, corn husks and waste rags have done the job. The Nouvelle toilet tissue website (slogan 'soft on you, soft on the environment') reports that English lords in times past advised purchasing inexpensive volumes of verse for the purpose, to be read first for educational purposes, and then used to wipe. Leopold Bloom's famously detailed bowel movement in James Joyce's *Ulysses* utilised the local rag he had been reading, a habit that endured until and during the Second World War.

That great innovation, the continuous roll of toilet tissue with tear-off sheets, was patented as long ago as 1870 by the Scott Brothers in Philadelphia. According to the delightful Bog Standard website (which promotes better toilets for schoolchildren), soft paper was introduced in 1932 but was unpopular at first, although it doesn't explain why on earth that should be. Now companies go to huge lengths (if you'll forgive the pun) to make and market to us the softest of soft loo paper.

Many millions of trees have been sacrificed to this end.

Rose George's exposition of global toilet habits contains the arresting statistic that Americans use an average of fifty-seven sheets of toilet paper a day. An equally arresting statistic is that when surveyed by a Dr J. A. Cameron of Oxfordshire in 1964, nearly all of the 940 men he examined had faecal debris on their underpants. George concludes that 'paper cultures are in fact using the least efficient cleansing medium to clean one of the dirtiest parts of their body'. And to cap it all, according to ethicalconsumer.co.uk, the British are possibly the world's most profligate users of toilet paper with the average person getting through 110 rolls a year.

If using paper is a must, surely it should be the recycled variety, or at the very least be certified as coming from 'sustainable' sources? Paper can be recycled up to six times before the fibres are too weak to cling together, but toilet paper can be made towards the end of this process. Yet of all the toilet paper on the market, in Europe only around 20% is made from 100% recycled fibre, and in the US only 2%, while sales of the super-soft kind increased 40% in 2008.

For toilet paper made from 'virgin', i.e. new, fibre, it should at least be FSC. Forest Stewardship Council certification (there are other standards, but FSC is the most well known) means that the pulp has an audit trail – it comes from forestry that has met certain good management standards, and certainly should not have come from illegal sources. Here, there has been greater market penetration than for recycled, which insiders in the retail industry put down to increased customer awareness and enquiries as to whether they use FSC pulp. It is often cited as a successful example of consumer labelling and subsequent market shift – a single issue, reasonably clearly defined and prominently labelled, prompting companies throughout the supply chain to ask for FSC pulp (or other standards of certification such as PEFC), which in turn creates an incentive for suppliers to get the certification. In 2006, it was estimated that the global supply of FSC-certified market pulp was approximately 4.7 million tonnes, amounting to 9% of world market pulp capacity. That may seem low but, worldwide, the number of FSC-certified forests is growing. In 2009, it stood at 100,000,000 hectares, and although that is still only 5% of eligible forests, it's a considerable achievement in the seventeen years since the system was started.

Perhaps the most important bit of 'non-information' from the consumer perspective is that *not* having certification doesn't necessarily mean that the paper hasn't been produced sustainably – it may simply mean that the producer hasn't seen any advantage in applying for certification, and that those procuring the paper in the supply chain haven't demanded it. If growers can't charge a premium for their product in the marketplace (as is often the case with organic food) then it may not be worth the bother. But then again, charging a premium might also lessen demand. It's a catch-22.

FSC also has a label that denotes recycled content, but not necessarily coming originally from FSC sources. There is a third label denoting 'mixed sources', which is quite a complicated category – it means the paper has a 'rolling average' content of 70% FSC-certified sources, but might also contain post-consumer recovered fibre, and fibre from 'controlled sources', which are not certified FSC but exclude 'unacceptable' forestry. These are all good things, although they mean three labels instead of the straightforward single one. Ironically, when I looked at the FSC website, the home page announced that the winner of its latest competition would win a year's supply of toilet paper. I was intrigued to know how much a year's supply was reckoned to be, although I accept that specifying the appropriate rate of toilet paper use is not really the FSC's job: they have enough to contend with in terms of improving forestry practice.

If we have to use precious fibres to clean ourselves, a better prospect might be to make them out of other waste products of various kinds. In some shops, I can buy toilet paper made from cotton, and organic cotton at that, using the waste fibres from textile production. That might work as a mainstream product if organic cotton was more widely grown, which would be a good thing in itself. US environmental organisation Worldwatch Institute reports that the three largest toilet paper producers in Japan use recycled wood pulp with 'washi' as a paper additive; washi is made from a variety of sources, including rice, hemp, bamboo and wheat. Toilet paper innovation does not seem to be going in quite the same direction in the UK. In the same store as the organic cotton rolls, I can also buy toilet paper with 'extract of cashmere'. I can't help feeling that this is unnecessary, but perhaps only as unnecessary as the papers with aloe vera and all kinds of scented, lubri-cating additives.

Which is best? The cheapest, greyest, recycled kind from a discount store, or the FSC branded? Some of the latest carbon footprinting exercises suggest that recycled is lower carbon, and not bleaching it saves on potentially polluting chemicals; on the other hand, we should be encouraging FSC. No one is making a definitive judge-ment, and the market is free to make toilet paper out of silk and satin, if that's what it thinks will sell. All this, to me, makes the case for limiting choice, and setting a stan-dard to ensure that loo paper, at the very least of all the paper products we use, is made from recycled paper, so long as it is made by a process that doesn't use more energy and therefore carbon. For an application that is very definitely the end of the line, it seems logical to use paper fibres that have already been recycled several times and are near the end of their useful lives.

Best of all is to use as little as possible. Japanese archi-tect Shigure Ban has designed a loo roll with a rectangular cardboard inner, the intention being to allow less 'spin'

so less paper is taken from the roll. That has the additional benefit of making it harder for children to playfully
pull out reams of the stuff and leave it all over the bathroom floor.

Too precious to use

Sometimes the straightforwardly 'green' choice exists, but
consumers haven't heard enough about it, or acted on
the information if they have. This, I contend, is the case
with compost and peat.

Peat bogs are precious places, with their starry mosses
and insect-eating sundews. They are a rare habitat found
in the UK, Ireland, the Baltics and parts of Spain, and
there are versions in tropical forests. They are a function
of very special conditions where waterlogged land
becomes colonised with mosses. As the mosses die and
are compressed by the water and moss above, they
become peat. Bogs are generally in windswept places and
their cushions of bright green, inviting moss interspersed
with deep channels of water make them treacherous
places to be, but also places of surprising beauty. They
are home to an impressive array of species, most of them
tucked up among the moss.

There is another reason not to disturb the bogs. Made
of compressed, dead plants, peat, like oil, is a carbon
store (and indeed is traditionally used as fuel). But if
drained and dug up, the peat oxidises and releases
carbon dioxide into the atmosphere. An estimated 3
billion tonnes of carbon is locked in UK peatlands, and
if lost as CO_2 would be the equivalent of twenty years
of industrial emissions. At present, the peat dug up in
Britain to make compost for growing plants releases
almost half a million tonnes of carbon dioxide a year –
the equivalent of the emissions from 75,000 households.
Drained peat bog can also spontaneously combust and

smoulder for years, as has happened in the Las Tablas de Daimiel national park in Spain.

Since peat renews itself at only 1mm or so a year, it is in effect a non-renewable resource. Gardening is a relatively new use – most potting composts used loam (sieved soil) until the 1970s. Peat is beloved of many gardeners, as well as commercial plant producers, for seed-raising and potting. I cannot share their enthusiasm. I find it a tricky planting medium, and having experimented widely, I much prefer alternatives such as composted bark or coconut shell (coir), or more recently composted municipal green waste. Even if I did think peat was horticulturally superior, I would rather have the peat in the bog than in my greenhouse.

Peat-free compost has been available since the 1990s. It has had high-profile advocates in the form of celebrity gardeners and wildlife broadcasters, as well as a campaign by WRAP, who want to see peat replaced by composted kitchen and garden waste, thus providing a 'closed loop' for organic materials from households back to households. Given this encouragement, how can gardeners, who might be supposed to be connected to the earth in various concerned ways, continue to buy peat? The answer, surprisingly, is that a lot of gardeners are not concentrating on being 'green' consumers. A National Trust poll revealed that only 29% of gardeners bought 'peat-free' brands, despite their being widely available, and 39% said they didn't realise that brands not labelled 'peat-free' probably did contain peat.

Partly this is a failure of labelling (labelling of peat content should be required) but it is also a failure of 'green' consumers to show sufficient interest. It is going slowly in the right direction – according to research for DEFRA the use of peat was down slightly in 2007 (from 3.4 million cubic metres in 1999 to 3.01 million cubic metres, still an area equivalent to 12,000 Olympic-sized swimming pools) and the proportion of peat in composts was gradually dropping (from 95% in 1999 to 72% in 2007). But still there is no rapid end in sight for peat in composts, so the green choice has not yet made sufficient difference.

Crisp choice

Sometimes the choices are ahead of their time. Here I put the carbon-labelled crisps. Some crisps carry a label giving the amount of carbon that has been emitted in the course of their 'life cycle'. But most don't, so using the label as a decision-making tool when buying your lunchtime snack is some way off. Behind the label is a process of 'carbon footprinting', which does a very thorough job of calculating the carbon used to grow and harvest the potatoes, process them into crisps, package them, drive them to the shop and market them. This has yielded some important insights – that a third of the carbon is in the packaging for instance. Walkers, who collaborated with the Carbon Trust to do the trial footprint, claim to have reduced the weight of the packaging used across the process, using lighter corrugated boxes, and also achieved a 4.5% reduction in emissions associated with producing individual packets. It also emerged that farmers were humidifying their potatoes, and because they were paid by weight this gave them higher value, but it cost them energy to achieve, and cost the crisp company energy to drive off the excess water in the frying process. A better way would be to reward the farmers according to how *low* they could get the moisture content.

All this is important, but in what way could the consumer exert an influence? Is the difference of a few grams of carbon on the label going to a) override other considerations, not least taste, or b) direct the market in a way that results in increased company activity on the issue? The label can only be done on the back of the 'footprint' exercise, by which time the company should be reasonably clear what needs to be done. It seems unlikely that consumer choice can add anything to the mix of drivers that then kick in, including potential cost savings for the company and government targets for carbon. And if there were no government targets for carbon, I wonder how many companies would be going through this exercise at all. Carbon footprinting presently costs around £30,000 per product, so the likelihood of it being applied in this form to all products in the near future is small – it is more likely to yield supply chain insights that a number of companies can act on, ideally together, to improve whole product lines.

That is not to say that knowing the carbon footprint of things doesn't have educational value, and even entertainment value. Mike Berners-Lee's very thorough book *How Bad Are Bananas?* gives us page after page of astonishing comparisons. For instance, if you cycle a mile powered by bananas you'll be responsible for the emission of 65g of carbon, whereas if you've stoked up on a cheeseburger it is 260g, about the same as would be emitted by an efficient car. That's just a bit more than an hour watching a 42-inch plasma-screen TV, or having a large cappuccino, but less than a pint of locally brewed beer drunk at the local pub, and only about a tenth of the footprint of a single rose grown in a greenhouse in the Netherlands, whereas a home-grown, organic rose is zero. All that should help you decide how to spend your Sunday. And the answer to the cover question? Bananas are quite good in carbon terms since they are not grown in energy-intensive hothouses, keep well on long journeys and come in their own packaging. But there is more to them than carbon – we need to consider whether wild land has been cleared to grow them and how much pesticides they need, factors that are hard to distil into a carbon footprint. Back to where we started – not enough information.

Where do we get our stories from?

Why is it so hard to know enough and make the appropriate judgements? For a start, as we've seen with the carbon footprints and also the paper example, there is no comprehensive labelling. Paper or timber without an FSC or other 'sourcing' label may be just as 'good' as FSC, it may just be that the supplier has not chosen to apply for accreditation. It costs money to do so, and it may well cost money to change the way the forestry is done in order to get the badge. There might not be sufficient benefits in terms of extra sales to justify the cost. On the other hand, paper without the FSC logo may be 'bad'. We just don't know. FSC is a market-driven means of businesses meeting consumer expectations, but those expectations are not yet sufficiently forcefully expressed to make the entire supply chain adopt, or plan to adopt, FSC or similar accreditation.

Likewise, the European eco-label is something that companies choose to apply for rather than something used in a comprehensive way. Here the judgements are even less clear because the label wraps together different kinds of environmental impact, so although purchasers know that the eco-labelled product is 'better' than other similar products, it is not clear without looking at the standards behind the label in exactly what ways they are better. As of mid-2010, only 1,016 products carried eco-labels out of the thousands and thousands of products on the market, and the product groups covered are an odd mixture – lots of detergents and a couple of categories of paper, alongside hard floor coverings and paints. The Department for Environment, Food and Rural Affairs (DEFRA) says that the label is not a very effective driver on its own and needs to be accompanied by 'other policy measures'.

Worse than no environmental comment at all, however, is the labelling that has no expert backing – green claims put on products by companies. Here I am thinking of statements like 'kinder to the environment', so vague as to be almost beyond investigation. There is a government-backed 'green claims code' that regulates such assertions, enforced by the Advertsing Standards Authority (ASA). This swings into action only if someone complains (and complaints are rising as the number of green claims rises), but the sanctions are minimal – being asked by the ASA to withdraw the ad, and being named and shamed on the ASA website. The consumer magazine and website Which? tested a few of these kinds of claims with an 'expert panel' early in 2010, and concluded that, for many products, what was claimed for them (for instance, greater 'biodegradability' of cleaning products) was actually not significantly better than for other similar products. The Which? research suggested that 80% of its readers didn't believe green claims anyway, I suspect because of their very vagueness and inconsistency, making the whole business of asserting and then policing green claims seem rather pointless and expensive.

Besides this partial labelling, misleading labelling and non-labelling, we have little to aid our choice of products. Few companies publish information about the sources and credentials of their products in an accessible way. The government's Advisory Committee on Consumer Products and the Environment (ACCPE), now disbanded, spent six years pondering how best to do this. Alan Knight, former environment director of Kingfisher plc (owner of B&Q) and now an adviser to Virgin Group, who chaired ACCPE for all of those six years, championed the idea of 'product stories'. These should not be lengthy life-cycle analyses, but a summary of the main interesting bits about products, which Knight believes is what consumers want most. If products could talk, he asks, would they be proud of how they are produced, sold and discarded after they are used? Or would they be embarrassed or ashamed? And who would want 'slash and burn' garden furniture, if that was the story it told about how the timber had been harvested?

ACCPE could have revolutionised our understanding of products, but the government's take-up of the committee's recommendations has been mixed. ACCPE's original mandate was to look into creating a UK eco-labelling scheme, but the committee's research showed that trying to reflect multiple issues on a single label was likely to fail, because of the lack of transparency of particular product impacts. Alan Knight went on to co-chair the Sustainable Consumption Roundtable, another government-spawned body, which provided further evidence that consumer labels had limited effect, and that the real drivers for change were regulation and companies' buying practices. The latter included the idea of companies 'choice-editing' on behalf of consumers and deciding not to stock products with high environmental impact; Wyevale and B&Q were among the companies at the time declaring that they would stop stocking the controversial and highly energy-intensive patio heaters. In terms of more comprehensive consumer information, the government did flirt briefly with the idea of a website holding data about products under the title of 'Environment Direct', but this got watered down to being general advice about what you, the consumer, can do rather than being a way to access detailed product information. The Roundtable also introduced the concept of product road-mapping, where industry sectors got together and agreed to look at the main environmental impacts in their supply chains and what they could do about them. DEFRA took up the recommendation to trail ten 'product road maps' to test the idea, hoping that other sectors would take up this kind of initiative. The clothing and dairy road maps in particular have been successful in generating debate about their supply chains.

It has taken longer to implement ACCPE's central message, that there should be a comprehensive approach to assessing, influencing and communicating product impacts, and a dedicated independent body to carry out this approach. It is now finally being adopted with the expansion of WRAP from looking at recycling of packaging to looking more broadly at products. WRAP is determined to make sustainable design the norm, but does not yet have all the right instruments at its disposal to make this a reality. The overwhelming majority of product stories are yet to be told.

In the absence of official ways of telling product stories, it is left to NGOs and writers to fill the gap. One of the best I have read is a hard-hitting little book from the Seattle-based organisation Northwest Environment Watch, who had the inspiration to examine the 'secret life of stuff' long before I did. Their book *Stuff* takes the reader through an average day, tracking the origins of the foods eaten at breakfast, through what it takes to provide the daily newspaper, the clothes you might wear, the bike, car or computer you might use, right through to the cola, hamburger and fries for lunch. These are wonderfully detailed, albeit in a US context, so the reader gets a flavour of which forest the newsprint paper might have come from (the Cariboo Mountains of British Columbia), how it was logged (by clear-cutting rather than by selective logging), which paper mill it went to, what kind of pollution the mill generates, and how much of the newspaper might have been recycled. Unfortunately, it does this so successfully in terms of bringing home the variety of issues and problems involved in everyday stuff, that the prescriptions for personal action at the end of each section seem unequal to the magnitude of the challenge presented. I suspect this is exactly what the authors intended – they do say at the beginning 'prepare to be unpleasantly surprised'. More recently, Annie Leonard's extremely thorough book *The Story of Stuff* also explores the production system from the U.S. perspective, and with the benefit of Annie's two decades of research, travel and activism. She points out alternative ways of delivering our consumables, and deals with the issues of toxicity at greater length than I've been able to here. Her book is also peppered with 'signs of hope', an important quality in what is otherwise a very sobering story.

The nearest UK equivalent I have come across is Fred Pearce's *Confessions of an Eco Sinner*, one of my main sources of inspiration for writing this book. Fred, a science journalist for many years and author of several books on the environment, also goes in search of the origins and impacts of the stuff around him. He goes to the gold mine that very likely produced his wedding ring, to the poor farmers who grew the prawns for his curry and the rich farmers who grew the cotton for his T-shirts. He sees mining, logging, farming and also recycling and disposal all over the world. He brings home the human impacts of all these activities just as forcefully as the environmental consequences, and finds reasons for optimism as well as plenty of things to worry about. He is more thought-provoking than judgemental, and his book should be required reading for anyone with any interest in the material world, and certainly for anyone with a role in making decisions about it. But it would be over-optimistic to think that every consumer would make time to read it and allow it to influence individual purchasing decisions. Of the many people to whom I have recommended it, some have found it transformational, others have simply not seen the relevance of it to their lives.

American author Daniel Goleman wants to take a step forward and achieve what he calls 'radical transparency'. A psychologist by profession and best known for his book *Emotional Intelligence*, Goleman decided to turn his attention to the subject of 'ecological intelligence' to see how easy it might be to make informed choices about products in the US context. He concluded that it is virtually impossible except in niche areas, and he is angry about 'green washing' that trades on a few supposedly green attributes of a product while failing to give a rounded account of its total impacts. In Goleman's view, if any product on the market has hidden environmental impacts, every time someone buys it they are in effect rewarding those impacts. In his world of 'radical transparency', companies would produce full life-cycle data for consumers to enable more ecologically intelligent decisions, preferably aided by ratings systems that give an instant, in-store indication of the good, the bad and the ugly. A US organisation has started down this road with an Internet service called 'GoodGuide' – a portal to a number of technical databases that claims to enable a product's 'backstory' to be accessed by consumers. Goleman foresees a world of active consumer-campaigners using social media to communicate with each other about positive product stories, as presaged by the growing number of sustainable product 'wikis' and blogs. He argues that companies should want to give this information to consumers, because consumers will appreciate and act on it. But nowhere does he suggest that it should be *required* of those companies. This makes it doubtful that information will be provided consistently and comprehensively, and that the majority of people will seek it out.

And even when in possession of a life-cycle analysis we have to be able to interpret the results, or someone has to be prepared to interpret them for us. I once ploughed my way through a document setting out the life-cycle implications of a Nokia 3G mobile phone. It was a very thorough and informative document, but at the end, the author concluded that the lifetime carbon emissions were equivalent to 'producing and cooking five cheese cream potato gratins'. Why that dish in particular as a basis of comparison I never discovered, but in any case it didn't help me much. How bad is a cream cheese potato gratin in terms of its contribution to climate change? And would sacrificing five of them, assuming you happen to be particularly partial to them, offset the bad effects of your phone? By the time you've got to the end of the questions, the answers seem meaningless. The only real comparison would be with every other mobile phone, provided the same methodology had been applied – then at least we could choose 'best in class'.

The green consumer as a limited force for good

So at the moment the 'green' consumer has a limited range of opportunities for exercising choice, and those choices influence outcomes in only limited ways, based on the prominence and clarity of information carried on labels and in supermarket or environmental campaigns. The remainder of retailers' products communicate nothing of their environmental impacts, good or ill.

Please don't misunderstand – I am not arguing that trying to be a green consumer is pointless. But we need to ask for something different to what is presently on offer. We need to ask for the choices to be more actively made on our behalf, in case we are unable or unwilling to make them ourselves, as is often the case. We saw in Chapter 8 that it is almost inconceivable that the world can support continued economic growth and increased personal affluence, on a global basis, without some kinds of limitations kicking in. Unless we willingly reconfigure our present ways of using materials these limitations may not be of our choosing. We must be prepared for products as we know them to undergo changes, and be prepared to accept that some things will disappear and others take their place. This could be a very exciting,

creative and enriching process for us all. I hope to convince
you of that in Part 5.

Before we move on to the new 'design principles',
however, I want to share three snapshots from far-flung
parts of the world, personal perspectives from friends and
colleagues, that illustrate the shadows and light of where
we are now.

11

Letters from the real world

Dear Julie,

The light bulb glinting in the weak sunshine shows that the away in 'throw away' doesn't really exist. Mystified as to how it got here, Mum and I spend the rest of our walk debating the light bulb's potential journey. With the light bulb sitting proudly alone and in one piece, fly-tipping seems out of the question and since we're walking an uninhabited stretch of coast, there's no chance someone casually dropped it. We're fairly sure it must've been washed up out of the sea, but watching the fierce waves roll by, that feels close to miraculous. More so when you consider where we live. The Shetland Islands are the most northerly part of the UK and it is a twelve-hour boat journey from the mainland. When someone threw that light bulb 'away', I doubt this is where they imagined it ending up.

The light bulb is, however, far from alone. For generations, driftwood was virtually the only thing washing up on our shores and, in a relatively stuff-free society, was a highly prized commodity. The Croft House Museum, an example of a Shetland house circa 1850, demonstrates well the sheer number of uses driftwood could be put to – skilled hands transformed driftwood into mousetraps, box beds, farm tools and beautiful chairs. Even the dividing wall between croft house and animal byre is crafted out of a sort of wooden patchwork, no mean feat in a place without trees. When my parents arrived in Shetland in the late 1980s, many islanders still had their own wood-pile laid up on the beach and it was treason indeed to poach from someone else's pile. Over the years, the wood-piles have disappeared, giving way to an endless jumble of plastics of every colour, size and kind imaginable, mirroring their increasing ubiquity in our daily lives. A potent combination of geography and violent winter storms now makes the ragged Shetland coastline the end of the road for a seemingly endless amount of lost shoes, discarded packaging and the occasional light bulb.

What makes marine litter so difficult to deal with is its creeping, chronic nature – always there but never top of anyone's agenda. By the close of the 1980s, Shetland had had enough. Much of our livelihood as a community relies one way or another on Shetland's stunning envi-ronment, which was increasingly threatened by the hazardous and unsightly bounty washing up onshore. Determined to take practical action, in 1988 the Shetland Amenity Trust set up Da Voar Redd Up (which means the Spring Clean in Shetland dialect). The idea is simple: over one weekend, volunteers collect rubbish from wherever it has accumulated – beaches, roadsides and other public places – and gather it together for disposal. From just 400 volunteers in 1988, last year's Redd Up involved 4,125 volunteers – roughly one-fifth of Shetland's population – who picked up 65 tonnes of litter.

Once collected, though, the disposal options for debris like my light bulb, and Shetland's waste in general, are somewhat constrained by our isolation. With the nearest land almost a hundred miles away, transporting waste for processing is just not economically viable. Shetland therefore has to be relatively self-reliant when it comes to waste management and utilise the capacity within the islands to reuse, recycle and dispose of waste.

This translates into unusual policy solutions. Shetland differs from many local authorities in that we do not rely on landfill as the primary destination for our waste.

Instead, we have a waste-to-energy plant which burns municipal solid waste (MSW) from Shetland and the neighbouring Orkney Islands, and powers the district heating scheme in our largest town, Lerwick. Currently 20,000 tonnes of MSW along with 2,000 tonnes of offshore oil-related waste are burned every year and the heating scheme is operating at capacity, serving about a thousand buildings. The throughput of the facility is expected to remain the same in the coming decades, although there is capacity to burn more.

Considered alongside landfill, incineration is undoubtedly a more favourable option, but it still remains the Achilles heel of Shetland's waste management strategy. Public investment of approximately £22 million in total was needed to build our 'energy recovery' facility and the accompanying district heating scheme. The pay-off has been twenty-six jobs, up to £3 million kept in the economy each year from fuel savings and a significant reduction in fuel poverty, as well as substantial engineering and plumbing contracts for local firms. The sheer scale of investment and the need for consistent levels of waste to maintain the performance and viability of the waste-to-energy plant make this no stop-gap solution but part of a long-term plan which not just permits us but actually requires us to generate a steady stream of waste. Tying the scheme to tangible social and economic benefits further complicates this by making the critical need to reduce waste, not just process it more appropriately, appear less desirable and less urgent than it really is.

This situation is undoubtedly peculiar and frankly a bit disappointing – ideally I'd rather not have people's livelihoods tied to my continued ability to generate rubbish – but berating my local council misses the point. Our reliance on incineration really just brings home the uncomfortable truth that Shetland is as much at the mercy of current product design and prevailing waste(ful) behaviours as anywhere else. Shetland Islands Council actually does a pretty amazing job. Their first priority is waste prevention, by ideally capping waste generation in the isles at 2003 levels by 2020, and they show an admirable willingness to investigate substitute sources of fuel for 'energy recovery' should MSW fall significantly. But Shetland has 'stuff' to burn and will continue to do because companies show an alarming lack of foresight and concern about the end destination of their products. Companies design in waste, individuals and local authorities just play with the few options they have to reuse, recycle and dispose of that waste.

Recycling in Shetland is also unusual in that recycling rates are not projected to increase much beyond current rates of around 23%. This is in direct contradiction to the Scottish government's zero-waste proposals which seek to recycle 70% of wastes by 2025. Nowhere better illustrates the government's inability to recognise that waste management needs are diverse and place-dependent than Shetland, where recycling approaches that rely on centralised processing plants and distant markets are unfeasible both economically and environmentally due to the distances involved. Our geographically dispersed population, across fifteen inhabited islands, similarly makes door-to-door collection fairly impractical on any large scale and harder to justify environmentally. That's not to say we can't recycle more, just that recycling needs to focus on local needs and opportunities. We have some fantastic recycling initiatives, frequently in the form of social enterprises, and current projects range from a local glass reprocessing plant, which manufactures construction materials, to the processing of scallop shells into horticultural products.

From making driftwood into mousetraps in the 1850s to a contemporary local business which makes teddy bears out of used Fair Isle jumpers, Shetland also has a wonderful tradition of reuse. Prior to the immense oil wealth which arrived in the 1970s and 80s, Shetland's economy centred on crofting and fishing, both of which remain strong in the islands today. Making a living from the land and sea in an unforgiving environment encouraged habits of thriftiness and making the most of what you could lay your hands on. Skip diving is an institution, to the extent my dad would steer us on Sunday-afternoon trips out based purely on the reputation of particular skips around the islands.

Waste isn't just seen as a necessary evil, it's very much seen as an opportunity. On a large scale, this is most evident in our pursuit of and success at attracting oil decommissioning contracts. Close proximity to many oil fields, decades of experience in the oil and gas industry and high-quality facilities make Shetland very attractive for companies seeking to decontaminate and safely dismantle and dispose of redundant offshore installations. At a community level, diverse reuse, refurbishment and recycling schemes all bring employment and economic opportunities which are hugely important in such a small community – waste management is seen as one way to keep jobs and people in the islands as Shetland looks to a future beyond oil.

Shetland is a remarkable place and has a similarly remarkable relationship with waste. From burning it to recycling it, reusing it to removing it from our beaches, we manage our waste in a way that departs from the well-trodden path of landfill and centralised recycling depots. Our geographical isolation does undoubtedly act as a constraint, but it also works overwhelmingly in our favour by forcing us to find new approaches which embrace our local situation. But alongside all of that, we know that, globally, designing out waste must be the top priority.

Dear Julie,

My family moved to America two years ago, and one of the things I found hardest to come to terms with was the incredible waste I saw almost everywhere I looked.

America truly is the land of opportunity. The opportunity to shop.

There are miles of shopping malls, strips, discount this, discount that, sales, offers, coupons, vouchers. When we first arrived it seemed to us that the whole country and economy was based on a huge money-go-round, and so much of this consisted of the typical American just buying 'stuff'.

Some of the discount stores are vast. Our local one is probably as big as a typical IKEA store in the UK, and we use IKEA-sized bags to carry our shopping home, the boxes and sizes of groceries are so big.

There is so much encouragement to buy big. A good example is at the cinema. You have a choice with popcorn: you can buy a medium bag (about as big as a large in the UK) for $4.68; or you can buy a large which is 50% bigger for just $4.98 – and you will get a free refill with that one. That is 200% more popcorn for just 6% more money. Coke is the same; burgers, clothes, clubs and life in general shout out to you more is better. No one needs that much stuff so inevitably much is left uneaten or unused. I heard once that near to 40% of all perishable food is thrown away untouched. I can certainly believe this.

When we looked at houses to rent, we were astonished at the volume of possessions people had in their homes – often the basements (as big as some UK houses) were full of stuff they were storing. I tried to imagine the life story of one of these items. Raw materials probably mined somewhere in Asia or Europe, oil from the Middle East, all come together and processed and packaged in China, and then shipped to America to be stored in a cardboard box in someone's basement – unused and unwanted.

US houses are often vast. These vast houses have many rooms. We know many people who have a full-sized sitting room that is just used for Christmas. Often there will be three family rooms – one for the family, one for just the children, and perhaps one for when guests or friends come round. Each and every room will be filled with furniture, TVs and all the normal stuff. And these rooms are big. Each and every room will be heated in the winter (which is very cold), and cooled in the summer (which is very hot) despite the fact that often these rooms are hardly used.

Where we live, you would probably need to clear snow from your drive four or five times a year. I use a shovel, it is good exercise, and cheap. Almost everyone else owns and uses a petrol-driven snow blower. Same thing with leaves in the fall – petrol-driven backpacks that blow the leaves off your garden (and onto the garden of your neighbour, who will then use a petrol-driven leaf blower to blow them right back at you).

A good friend of ours once told us that life is a game, and in America he who has the most 'stuff' when he dies is the winner.

Americans love their coffee. It's a stereotype, but true, that Americans walk around with huge cups of coffee in their hands. I once saw two people out on a hike, four miles from the nearest car park, and God knows how much further to the nearest store, with a cup of coffee each from a doughnut store.

When I went climbing in the Catskills, there was a park ranger checking everyone had the correct pass, and bizarrely, handing out cups of coffee. There were no bins available.

Here is the thing with drinking so much coffee: Americans are estimated to throw away 25,000,000,000 Styrofoam cups each year. These cups could circle the planet seventy times per year. Worse, Styrofoam does not decompose.

Americans also love soft drinks, which they call soda. The US airline industry discards enough aluminium cans every year to build nearly fifty-eight Boeing 747s. That's a lot of soda cans. Every soda can thrown away might as well be a third filled with petrol because that's how much fuel is needed to make a new can from scratch. A typical American will use a 100-foot Douglas fir tree per year in paper and spends up to eight months of their life opening junk mail.

Yes, there is recycling over here. Though I've noticed that in my neighbourhood people seem to care less about recycling than in the UK. And yet the collection service is good – they come every two weeks and collect paper, cardboard, metal, plastic and glass. Waste collection and recycling provision varies wildly from state to state: Albuquerque = no recycling; Portland, Oregon = lots.

It is also very hard to find recycling bins away from the house, much less common than in the UK. When we stayed in the National Parks last year, it was next to impossible to recycle a Coke can. We were amazed and disappointed about this – if there is one place in America where this message should be shouted out loud, it's in the National Parks. Plus, although there were lots of education and ranger-led activities for children in Yellowstone, we saw no mention of the environment, climate change, recycling or alternative energy sources, and given that it is the most famous and visited geothermal location in the world, this was very disturbing. We felt it would have been a great statement if the power for all the visitor centres had been harnessed from geothermal power, or at least some token display explaining the potential. There were many pointless souvenir stores, but we saw not a single exhibit on anything green. A clear sign that this is depressingly unimportant to many Americans.

Oddly, it seems that there is less street litter here than in the UK. Litter in America tends to be on a huge scale – buildings and machinery left to rot rather than cleared. Perhaps as space is not an issue, often it is cheaper to leave things and start again from scratch, but it does little for the scenery.

Overall, America is by far the largest producer of waste in the world on a per capita basis, and the demand for cheap goods from China is a huge factor in the nightmare that is developing over there.

Good luck, world. Let's hope American ingenuity can rise to the challenge.

Dear Julie,

I want to tell you about Kamikatsu. It is a small town on Shikoku Island, southern Japan, with a population of 2,000 people, made up of 800 households. The area is mountainous, and more than 85% is covered in forest. There are about 55 settlements dotted around at altitudes between 100 and 700 metres. Kamikatsu is a member of the group of 'the most beautiful villages in Japan'.

In September 2003, the town adopted a 'Zero Waste Declaration' to reduce the amount of waste incinerated or sent to landfill to zero by 2020. The town wanted to say 'no' to high dependence on incineration and 'yes' to a more positive goal – the people of Kamikatsu wanted to set an example.

Lack of space for landfill on the island means that incineration is the waste management strategy of choice. Seventy per cent of municipal waste in Japan is dealt with in this way. In Kamikatsu, waste was burned in the fields until 1998. Backyard burning used to be common but is now perceived as unacceptable and is reducing. Instead of increasing the amount sent to incinerators, Kamikatsu wanted to revive the principle of mottanai *(which roughly translates as 'it is a shame for something to go to waste without having made use of its full potential'). Japan was traditionally not a throwaway society. The hope was that the success of Kamikatsu would help make 'zero waste' a nationwide movement.*

The zero-waste initiative has been helped by national legal requirements to sort waste into categories, recycle packaging and recycle waste electronic goods. Gradually, the costs of collection and sorting waste are transferred to both householders and businesses. These requirements come from the overarching Basic Law for Establishing the Recycling-Based Society, enacted in 2000 to promote a resource-recycling society. But what is happening in Kamikatsu goes well beyond this.

On a household level, residents of Kamikatsu have been required to separate their waste into thirty-four separate streams since 2001. The waste is separated at the public 'Garbage Station' rather than through kerbside collection, since home collection is not economically viable in a town where the houses are spread out on mountainous terrain up to 700 metres above sea level. Therefore there is a heavy reliance on residents to recycle, despite the inconvenience of transporting their own waste. Recycle Kamikatsu, a volunteer group, transports the waste of senior citizens without cars. There are no legal requirements for recycling (as opposed to sorting), so it is an entirely personal responsibility. This is an unusual state of affairs as most other case studies seem to emphasise ease of kerbside recycling as a major instrument for success in recycling, and Kamikatsu has bucked this trend. Some people used to throw their rubbish onto their own mountain, but they are few as a result of the town's patrol.

There are thirty-four separate categories of waste because the town works on the assumption that the more materials separated at source, the more can be easily transferred to recycling processes. There are established markets for aluminium cans, steel cans, cotton, cloth and milk cartons, that are all sold to recycling companies at a profit for the local government. Collectors have also been found for such items as disposable chopsticks. Nappies are one of the few things going to incineration. Recycling of plastic products is subsidised by the Japan Containers and Package

Recycling Association. All other materials recycled (cardboard, newspapers, used kitchen oil) are recycled at a cost to the municipal government. Demand from China has developed the market for recyclable materials; however, this raises concerns about the dependency on China for the continuation of recycling in Japan, along with the environmental impact of transporting materials to China and the lower health, safety and environmental standards there.

In April 2005, a Zero Waste Academy was established to promote awareness about zero waste, and make policy proposals. The main activities of the academy – a small operation with a few paid staff and volunteers – are to manage the Kamikatsu zero-waste programme and to disseminate knowledge and expertise on the strategies. The academy holds seminars and encourages officials from other municipalities to visit Kamikatsu to consider the zero-waste goal for their own area. A magazine is published and distributed. In 2005, 3000 people visited Kamikatsu through the Zero Waste Academy's knowledge-sharing initiatives.

The result of all this effort is that, in 2005, recycling of household waste reached approximately 80% in Kamikatsu. This figure includes estimates of home composting, but even excluding home composting the figure is about 70% (based on 2003 figures). This is quite a staggering achievement considering the lack of legal measures in place (other than the requirement to sort) – there are no landfill bans and no financial instruments (such as landfill or incineration taxes) to add weight to the voluntary initiatives. The total 'residual' (not recycled) waste per person for our citizens is approximately 268kg per year, less than a third of that of San Franciscans, for instance. Now the challenge is to tackle the remaining residual waste. I know this is a tough challenge, as the remainder represents products and packaging that cannot be recycled by the municipality. This is a question of improving the design of products.

The mayor of Kamikatsu, Kazuichi Kasamatsu, believes that a law is necessary to require manufacturers to collect all discarded products and reuse the resources within them. He believes that there should be more emphasis on producing products that are less wasteful. Government and manufacturers need to work together and Mayor Kasamatsu has begun lobbying the financial sector and the Ministry of the Environment for waste reduction measures.

Home composting is an important means of waste prevention and there is a massive 98% participation rate in home processing of organic waste. The municipality does not collect organic waste but instead subsidises purchase of electric composting machines. These are like very large bread-makers with a heating element and a mechanism to aerate the compost, although the energy these require may sometimes outweigh the environmental benefit of home composting. Half of the households in Kamikatsu have taken advantage of this scheme. Most people are able to compost at home as they have their own gardens, and this makes it a lot easier for waste to be diverted from landfill in this way than in urban areas.

I am hopeful that the concept of zero waste can be exported throughout Japan. I am sceptical about the practicality of thirty-four separate waste streams in city areas, but outside cities it could certainly spread. Zero Waste Academies are needed nationwide to enable easier access to vital information and networks. Kamikatsu is too far removed from the more populous areas of Japan to be able to have the impact needed to bring zero waste to the cities. The key is the impetus of the mayor and officials, and other municipalities in Japan have already put the first elements of a zero-waste approach into action. For example, the mayor of Hino City (a suburb of Tokyo) implemented a 'No Waste' campaign in 2000 to tackle an area with the worst recycling rate in the Tama region of Japan. Through variable charging, home composting and separation of waste into nine separate categories for recycling, the municipality has made significant improvements. At the moment I am working in the government of Hayam town, an hour's travelling time from Tokyo. I am pleased to say that the town has set itself a challenge of reducing waste to landfill and incineration by 50% by 2014, and zero waste by 2029. So zero waste is spreading to urban areas. We think that Kamikatsu has helped lead the way to a better future for all Japan, and we hope the zero-waste message will spread internationally.

Part 5
Designing the Future

12

Where do products come from?

In Part 2, we deconstructed the things that make up our lives to think about their constituent materials, and their place in our current 'linear economy'. In Part 3 we saw that we might hit environmental limits of different kinds. In Part 4, we saw how we have little option but to consume (although we have choices about how much) and also how today's product choices offer little scope for even the most dedicated green consumer to lead a change away from the linear economy. Now we are going to look at how we might design things differently.

To get to the heart of what is needed, we have to think more creatively about how materials become products, and then become 'waste'. We have to think about the whole chain of production and consumption in which they, and we, are embedded – where they come from, how we use them and where they end up. Before humans came along, materials sat around in the ground, or as animals and plants, or just as chemical compounds. All were blissfully unaware of their 'usefulness' – they just *were*. It was humans who created uses for them, and now we take most of those uses for granted.

In 2009, I was delighted to discover that in the bowels of King's College London, there is a 'materials library' – the project of materials expert Dr Zoe Laughlin. Zoe's mission is to put on display in the two small rooms allotted to her by the university as many materials as possible, and illustrate their potential. I had an entertaining hour looking at Zoe's tuning forks made of various materials, from metals and woods to plastics and glass, and listening to their diverse 'notes'. I felt the weight of different metals, each of them rendered in exact 2cm cubes, from the featherweight feel of aluminium to the astonishing density of tungsten. One cube was a single unusually large crystal of the mineral pyrite, often confused with gold.

But beyond these curiosities, what is the relevance to us of materials? Only really their use in products. So Zoe's materials get even more interesting when she demonstrates that things taste better off a gold spoon rather than a silver one, or shows me the nickel 'super alloy' which is the only material strong enough to make jet engine turbine blades. Or demonstrates 'silly putty', a viscose elastic polymer (plastic to you and me), which is liquid enough to flow rather disconcertingly off the edge of the table if you leave it alone, but strong enough to resist knocks when used in protective sports gear and motorcycle jackets. I was also allowed to hold a magical blue cube of silicon-based 'aerogel', direct from the research labs of NASA. At 99.8% air, so only 0.2% 'stuff', it is the world's lightest solid. Zoe explained that it was designed to catch the dust in comet tails for analysis (some of it has made a 3-billion-mile journey through space) and owed its blue colour to the way in which light passes through it, making it blue for the same reason the sky looks blue.

Where do products come from? Chance discovery – someone at some point discovered the incredible usefulness of something as humble as beeswax, for instance. But also deliberate design, like the aerogel. Most things around us will consist of age-old discoveries overlaid by modern techniques, a shell of sophistication coating a dollop of ancient serendipity. So my coffee-shop coffee is a drink that the Yemenese and Ethiopians knew how to brew, at least five hundred years ago, but here it is delivered through an espresso machine invented in the early twentieth century and now utilising the best stainless steel, aluminium, copper and brass to good effect. The cup is ceramic, not far removed from those used by humans several thousand years ago, probably discovered by the chance sun-baking of clay, but the brand logo transferred onto it uses techniques first invented in the 1750s. The little plastic pots of milk and the printed, metalised foil wrappers on my biscuits represent the ingenuity of the modern polymer and packaging industries. Sometimes these industries come up with what they were looking for with their chemical stews, but sometimes they come up with something entirely different, which nonetheless might find an application. According to American sociologist Harvey Molotch, that was how Post-it notes were invented. Experiments at the 3M corporation came up with an adhesive so weak it was presumed useless, but a 3M engineer called Art Fry used a piece of paper coated with it as a bookmark in a book he lent to a friend, and the book came back with the paper stuck on the cover and a note written on it. The rest is multi-coloured, slightly sticky history.

Humans are constantly inventing, but the majority of inventions never go anywhere. According to one estimate, of 30,000 new consumer products launched worldwide every year, around 90% fail. Many things determine whether a technology or a particular product flies or fails, but the underlying requirement is that they find a place in the systems we already have. Harvey Molotch uses the example of a toaster: it is not just a question of having the raw materials available (different metals, heat-resistant plastics), but also of having an electricity supply at an affordable price, houses with sockets at table level and, most important, a taste for toast (Brits and Americans do, Italians apparently don't). Molotch calls these systems 'lash-ups' – a variety of things come together at a given time and place to produce the pattern of stuff around us.

This might seem self-evident in many ways, but it reminds us that living with better products is not just about individual product design or brand-new technologies (although these might be important), but about altering the 'lash-ups'. To do that, Molotch reminds us, we have to thoroughly understand how they work – how interacting factors such as prices, preferences and cultural norms come together before even thinking about the conscious 'design' process. So it's no good trying to promote electric cars if there is nowhere to charge them, or people don't like the sound the engine makes. We need to understand whether people want green cars that shout 'green' or ones that are indistinguishable from conventional cars. We also need to know what they are prepared to pay for them, and for the electricity needed to charge them, and whether they believe 'green' claims anyway, or indeed whether they even think that there is a problem that 'green' cars can solve. What story are people telling themselves when they decide they do, or do not, want an electric car?

It is tempting to cut through these complexities and see price as a main determinant of choice. We know that price is a big factor in the products we all buy, but we also know that people are prepared to pay different prices for similar things for what one might think are strange reasons – familiarity, fashion, a trusted company name, the recommendation of friends. Economists call this 'price elasticity' – we allow the extent to which price puts us off buying something to be stretched by a range of factors, only some of which are about real need. Conventional economists, who hold that people are rational, selfish and independent, might have trouble explaining this phenomenon – why pay more for essentially the same thing? Behavioural economists, on the other hand, understand it perfectly well, as we saw in the chapter on consumption. It is born of basic human instincts to stay in groups, seek the approval of others, display status and store up stuff in case of future shortage. Hence the tag 'conspicuous consumption' – consumption designed to say something to others. So the prices we pay for products vary depending on what we think and feel, sometimes changing from one day to the next. It is another way of saying that what counts is the way the whole picture fits together – the 'lash-ups'.

The trouble is that few bits of this tangled web are either consciously or unconsciously catering for environmental pressures. Prices do not on the whole factor in environmental damage, so the prices people pay cannot, except in very particular circumstances (such as paying an environmental tax), take care of that damage. At the same time, until very recently, designing products to address environmental concerns has not been part of businesses' perceptions of what constitute our most important needs and desires. On the contrary, as we saw in Chapter 10, catering for what has tended to be construed as a niche concern through limited product labelling has to some extent let companies off the hook about everything else. The consequence is an economy comprising a great many environmentally unfortunate 'lash-ups'.

I am an unwilling participant in this web, just like many others. Even with my strong environmental commitment, I have held off buying a photovoltaic panel to generate electricity at home (despite thinking that it is one of the better long-term energy bets) because it is a big outlay, I can't be sure how long it will take to get a payback and I don't want to look silly, I don't know many other people who have them and can therefore vouch for them, and I'm not quite sure how it would go down with the neighbours. I haven't been bombarded with sales materials from well-known companies telling me how easy and cheap it is to get one, and I haven't seen any messages from people I consider reasonably authoritative (DECC, or the Committee on Climate Change) assuring me that they will support and applaud me in my brave decision. I am not (as yet) paying insupportable sums of money for gas and electricity, which might be the case if the environmental costs of those fuels were factored in, through a carbon tax or other mechanism. All these are enough for me to hold off signing the cheque and then persuading my husband to get on a ladder and scrape the ivy off the roof (which we ought to do anyway) in order to install one. In any case, that means dislodging the cheerful colony of sparrows whose tweetings awaken me every morning in summer, and I like that.

I am comforted by Harvey Molotch telling me that I shouldn't blame myself for all this; it is a consequence of a long process of industrial evolution that can only be changed by some definite new rules. No one set out to do the amount of environmental damage we have done, certainly not us as individuals; it has come as a by-product of a market-based system and of the 'lash-ups' we've jointly created. It is now a considerable task to unravel. As Molotch puts it, 'A corporate conspiracy would be better news.'

We know that more than technology and prices are needed – we need whole-scale cultural buy-in to the project of building Evie and Ed's world. Having rehearsed the problems that seem to be around every corner, I want to show that there are solutions, that there are ways to get that buy-in, by drawing a vision of what the new 'lash-ups' could look like, and how they would add up to a more creative, satisfying world.

Design really means 'intent'. If our intent changes, then the practical manifestion of that intent in the shape of well-designed solutions will follow, be they products, materials, agricultural production or any other system. Design is not just about people labelled 'designers', and designers are not just people who work on frocks; they work in every kind of industrial production, from aircraft to bidets and from bleach bottles to X-ray machines. Designers may be the people responsible for the technical qualities of a product, or they may be the more 'arty' end who take something prosaic and make it into a thing of beauty – one definition of good design is that it does a job well and *looks* like it does. But in all cases, designers work within a brief, rather than setting the brief themselves. Imagine a world in which the brief always features the environment.

So this is the brief – and everything subscribes to a set of six principles:

1.   **Everything sustainably sourced**: all materials are sourced according to strict conditions on managing their human and environmental impact – this includes timber, minerals, textile fibres and even water.
2.   **Everything designed for recovery**: all products are designed for their materials to be recovered and recycled.
3.   **Nutrients cycled**: valuable nutrients from food, animal and human wastes are returned to the land to minimise the use of artificial fertilisers; as are biodegradable materials such as timber, paper, textiles and plant-based plastics when they can no longer be recycled.
4.   **All energy renewable**: but demand for energy may also need to be kept within limits.
5.   **Stemming the flow of stuff**: there are strategies to reduce the flow of resources through the economy, including keeping products in use longer.
6.   **Care with new promises**: new technologies are very carefully scrutinised – developments such as genetic modification and nanotechnology have to subscribe to the principles above, and not add any new problems.

There are many things going on already that start to give life to these principles – the trick is to bring them all together. In the next chapters I'm going to illustrate a few of the ways in which the principles might be translated into practice. In the last two chapters I'll talk about how that good practice might be ensured through the way we seek to condition the marketplace.

13

Everything sustainably sourced

This is the principle that ensures that any material is considered right from the moment that anyone thinks about extracting it, whether it is timber, metals, fibres or any other resource taken from the natural environment. Here are a few examples of how that might work.

Stewardship for metals

As we saw in Chapter 10, ensuring that consumers have a way of accessing a product's 'story' is a necessary part of improving that story. This approach has transformed parts of the timber industry. The Forest Stewardship Council certification and other similar schemes provide a means for buyers to get some reassurance that good standards have been met, and provides an incentive to sellers to meet the standards in order to access markets where buyers are becoming increasingly discriminating. More recently, this principle is being examined in relation to metals, with the idea of certifying particular mining operations as having adhered to good social and environmental practices.

Traceability and certification are harder for metals than for timber. Pieces of timber can be marked as they move from forest to suppliers to manufacturers; metals are sent from mines to processors, often from many different sources at once, changing their form from ore to metal to finished article. Ensuring that any given product is made from the metal from a particular mine is therefore difficult. When the Eden Project wanted a 'sustainable' choice for the roof of the education centre ('The Core'), the team chose copper, for its adaptability to the complex architecture, its durability and its beautiful colour. Some felt that this should have been 100% recycled copper, but the team wanted to make a point about ensuring the sustainability of virgin supplies, which is just as important as encouraging greater recovery. This meant being confident that the metal had come from a mining operation that could be proud of its practices. Rio Tinto offered copper from its Kennecott mine in Utah, where the company's efforts to clean up pollution and restore previously mined land have led to its being regarded as one of the most responsibly managed mines in the world. However, keeping the Kennecott copper separate from other sources meant that the German manufacturers of the copper sheets had to stop their production process and clear all other feedstock from the factory before processing the Eden metal.

The Eden team is clear that this 'protected stream' way of ending up with metal from a certain mine is not practical or desirable over the long term. The Core roof was, in effect, a stunt. What it points to is the need to ensure that all the players in the supply chain, from mine to finished product, want to be part of the 'good practice' story, and will influence each other so that eventually all raw materials can be certified. This approach is being pioneered by the Responsible Jewellery Council (RJC), which acts as an umbrella for companies who extract metals, work them, and design and sell jewellery. The RJC system, launched in 2009, certifies members' business practices, so that if you buy from an RJC company you know that they have adhered to some important standards – for instance, that revenue from diamonds is not helping to fuel conflicts, or that gold has come from Fairtrade accredited sources. It does not as yet certify the actual product, or track its movement through the supply chain to ensure that it is 100% 'good', but as of May 2010 it is consulting on ways to do that.

The idea of starting with jewellery was inspired, because it is something of high personal significance and durability for the buyer. No one wants to think that their wedding ring is a symbol of pain and pollution as well as partnership. There is a long way to go before this approach can be applied to less visible uses of metals – responsible nuts and bolts, for instance – but the model is there to be developed.

There are other sources of metals that perhaps we don't consider enough. In 2009 there were rumours, never substantiated, of the Japanese filling submarines with 'rare-earth' metals and cruising around their coasts with their precious cargo. But no submarines are needed – much of what is around us is a very useful metal 'stockpile'. The Second World War practice of collecting tools, pots and pans and even jewellery to help the war effort showed that at least some of this 'stock' can be pressed into use if needed. Today we have plenty of discarded appliances and gadgets lying around that could yield up metals with the right recycling technology. By the same token, what looks like huge carelessness by allowing financially and ecologically expensive metals to end up in landfill sites might turn out to be more far-sighted than it seems. Compared to some ores, these sites are relatively concentrated sources. In 2008, at the first international congress on landfill mining, scientists earnestly considered the engineering and the economics involved in recovering useful amounts of metal and other materials from the world's waste tips. They concluded that although not all landfills held significant goodies, there were no serious technical barriers to recovery and, with ballooning commodity prices, the economic case was looking strong. What we used to consider waste could become a 'sustainable' source.

Better textile fibres

As we saw in Part 2, textiles are more complicated than paper, metals and plastics. The stuff that makes them is more varied, and is more mixed together. There is an even greater range of implications, from all stages of their 'life cycle' – where fibres come from, how they are made, how they are used and recovered. In addition, clothes, like jewellery, are more personal than other material possessions; they speak powerfully of who we are or who we'd like to be.

In Evie and Ed's world, fibres are chosen to have the lowest possible impact. It is a world in which wool has a renewed place. At the moment wool consumption worldwide is on a steady decline, displaced by our tendency to keep our homes warmer, and replaced by synthetic fibres, which are lighter and easier to dry but use more energy in their manufacture. Wool has considerable advantages. It is one of the few fibres yielded by animals that can be produced on a large-scale worldwide, and it has the merit of being a 'co-product' of meat. Sheep will cope with low-grade land not suitable for other forms of agriculture, so it need not compete with production of food crops. It has the disadvantage that processing wool for spinning takes a lot of water for cleaning and 'degreasing' the fleeces – removing the lanolin that is the sheep's natural water-proofing agent – but this could still make it a 'sustainable' choice in the wetter parts of Europe. Wool is also relatively hard-wearing, can now often be machine-washed at low temperatures, and is completely degradable at the end of its life as long as it is not mixed with synthetic fibres.

What future for cotton? Conventional cotton with its

high consumption of water, fertiliser, pesticide and energy is very likely something we cannot afford. Organic cotton needs much less pesticide, but has a lower yield per acre, so it risks competing with food for available land. On a global scale, we may need to live with decreasing quantities of cotton, which means that it becomes much more of a cherished, luxury fabric. This would take cotton back to some of its most important roots, as produced by artisan weavers, printers and dyers in the India of the seventeenth century, the ornately decorated and precious cotton 'chints' that so influenced English design (as chintz). Such fabrics were treasured around the world in a way that makes today's £2 throwaway cotton T-shirt a sick parody of the trade.

Organic certification of wool, cotton and alternative fibre crops such as flax (linen) and bamboo is the nearest thing we have to a 'sustainable fibre' accreditation scheme. It is not hard to imagine this being expanded and eventually becoming a criterion for those buying raw fibres, in the same way as is being developed for timber and metals. Such a scheme could aim to tackle issues from overgrazing by cashmere goats to competition of fibre-growing with food supply, and give authoritative pronouncements on the merits of all kinds of emerging fibres such as soybean protein and synthetic silk. A start has been made by Dutch organisation Made-By, which produced a proposed index of sustainable fibres (up to the point that they are spun into yarn) in 2009, although this sparked considerable controversy in the textile industry. Watch this space.

A way with water

Water is another area where a 'good stewardship' approach, backed up by certification, might help to condition how products are made. A start has been made on 'water footprints', trying to calculate how much and what kind of water is involved in making, using and recycling stuff. As with carbon footprinting, this is the first step towards establishing benchmarks, but for water it is much more complex to decide what is 'good' or 'bad'. For carbon dioxide (and greenhouse gases more broadly), the objective is straightforward – reduce emissions, however and wherever they are generated. Those grappling with water footprints point out that to do so sensibly, the *type* of water has to be somehow reflected. So we can think of 'green' water (what falls on soil as rain), 'blue' (what is in aquifers and rivers) or 'grey' (already in the treatment system – used as drinking and flushing water and reclaimed through sewage treatment works). And then there has to be some assessment of how abundant or scarce these types of water are, on a regional, and probably even seasonal, basis.

Water management specialists in the Netherlands have done some of the most detailed work on this area. They recommend the identification of 'hot spots' in the supply chain, just as we saw for the carbon footprints. So overuse of 'green' water (rainfall) can risk loss of wildlife – for instance, use of rainfall for jatropha production (which is a biofuel crop) makes it unavailable to native species of plants. In a different context, use of river water in California's Silicon Valley, where huge amounts are used in manufacturing silicon chips to keep the facilities ultra-clean, is at the expense of users upriver. And 'grey water', i.e. water that has already been through the water treatment system, can be compromised by excess nutrients from agriculture, as we saw from the earlier discussion of phosphates. Once these hot spots are known, they can become targets for companies' own actions, for government measures, and (if the good and less good uses of water are communicated effectively) potentially provide consumers with a way of choosing better products. For instance, if we do want to make a big shift towards plant sources of energy, the results of this kind of study can be used to select the crops and countries that produce bioenergy in the most water-efficient way.

Given the variation in water availability and use, setting standards will be complex, and communicating to consumers through a label is inevitably challenging, even more so than with carbon. Two UK NGOs, Sustain and the Food Ethics Council, have looked at water footprinting and labelling in relation to food. They saw weaknesses in the footprint approach, including the fact that measuring the *amount* of water used does not take account of how efficiently it is being used, or take account of social and ethical issues such as access to water, or say anything about how water that is 'saved' from the supply chain might be used elsewhere once a hot spot has been addressed. They thus favoured a 'water stewardship' approach that scrutinises how well companies who use water are managing water issues in the round, rather like the FSC scheme for timber. They also recommended that as standards of any kind are developed, they should be consistent across industrial sectors, including textiles, energy generation and mineral extraction as well as food and drink. Only in this way can we get a handle on the 'good' and 'bad' products in relation to water.

Substitutions – designing away from problem materials

If specific materials become unavailable or undesirable, for whatever reason, we will simply have to find something else for the job. Designers are endlessly creative but sometimes they have an uphill battle to make change happen. Take the Mosquito aeroplane, nicknamed the 'Timber Terror'. Rejected by the RAF initially in favour of heavier, slower, metal planes, it ended the war with a lower loss record than many other models thanks to its more nimble performance. What is important is conscious design to get the right substitutes, not more damaging ones. Replacing concrete is one of the areas that needs careful thought.

Concrete or wood?

Such incredible quantities of concrete are used globally that it is hard to imagine what could be done about it. Demand is dropping in developed countries as building programmes slow down, but is rising fast elsewhere, and China now uses half of the world's supply of cement. It should be good news that the raw materials for concrete are plentiful in the earth's crust (limestone, sand and gravel with a bit of iron and aluminium), but the fact that they are mundane, abundant minerals means we tend to think very little about the damage caused by quarrying and sea-dredging. Twenty per cent of UK aggregate comes from the sea, and aggregates are sucked up using a giant Hoover – it does serious immediate damage to any sea life that it encounters, and also carves deep trenches through the seabed, which can worsen coastal erosion. The best thing is to use waste materials from other forms of mining, such as the china clay waste used in Cornwall. There are also materials from dredging that is carried out as a necessary part of coastal management, or 'secondary', recycled aggregate from demolition of existing buildings, or even recycled glass.

Aside from this, it is tempting to conclude that concrete's main problem lies with the large amount of energy needed to produce it, and the main focus should be on providing cleaner, more 'renewable' sources of energy. But that ignores another resource consideration – concrete is strong when compressed (as under the weight of a building) but cracks easily when put under tension i.e. stretched or twisted in any way. That is why most concrete in buildings is reinforced with steel rods as well as needing steel superstructures, and also why building booms see consumption of concrete and steel rise in tandem. Newer developments such as using fibreglass for reinforcement cleverly get around the need for metal, but don't do much for the recyclability of the material. More promising is the use of organic fibres instead of steel.

Is it possible to substitute concrete with other materials? Concrete does things that no other material can do: it is a good way to provide solid foundations for a building; it is important for 'thermal mass', i.e. as a heat sink to keep buildings warm; and although not as long-lasting as stone, it is relatively durable compared to timber, especially in wet climates. Wood has recently been positioned as a more 'environmentally friendly' material for housebuilding than concrete, mainly on the grounds that it is from a renewable source. Whether it really is 'friendlier', however, rather depends, as with paper, on where it's taken from – if grown as a 'crop' on already clear land, possibly, but if taken from virgin forests it may suffer from all the same problems as timber taken for papermaking. Given the quantities of timber already taken, and the challenge of doing this at a renewable, non-damaging scale in future, it seems hard to imagine that a big swing to wood is the right thing.

That's what I thought until seeing one of the exhibits at the Eco-Build exhibition in London in 2010. Flanked by soaring panels of honey-coloured wood fixed together in giant shapes like a three-dimensional jigsaw, the commercial director of a large Austrian timber company explained that 'cross-laminated wood' (layers of spruce laid at right angles and glued together to form super-strong sheets) can allow timber-walled buildings to reach at least nine storeys, possibly in future fifteen. Their timber is taken piece by piece from forests in Austria, where annual growth rates of the forest that naturally cloaks the mountain areas exceed the amount taken by more than 30%.

There is so much wood in Austria that the waste from producing building timber is burned in domestic and industrial boilers, and provides some 12% of the country's energy. The other advantages of cross-laminated panels include quicker building times, better retention of heat, and better resistance to earthquakes (they absorb shock more effectively than concrete). According to the company, the glue is solvent-free and formaldehyde-free. This means that if the wood is incinerated it produces no more residue than unglued timber. Putting all that together, there seems a strong case for looking at how far timber could replace concrete on a local basis, particularly where wood is abundant. As with many other materials, the need is for the right material in the right place.

Could we make buildings out of alternative materials? Concrete is a clever composite of different materials, so old that we have ceased to think about it. But there are newer composites in view that actually utilise even older materials and techniques. Just down the road from where I live there is a business park with buildings made from hemcrete, a mixture of lime and hemp waste (the plant that gives us both ropes and cannabis, although they are very different varieties). Lime is an ancient building material and relatively environmentally friendly, because of its abundance and the fact that it doesn't need cooking.

This kind of material may not rapidly displace our concrete jungles, but it may offer valuable lessons for sustainable building systems more generally. Between 'modernism' as an architectural movement (which brought us the clean-lined, cubist, mainly concrete-and-glass buildings that signified a break with the past in the early twentieth century) and the fad for high-rise which symbolises economic growth in capitals around the world, architects have delivered urban spaces that many would argue have lost touch with basic human needs. They require vast amounts of energy to heat and cool, they use materials that are hard to reincorporate into future buildings, and they provide indoor environments that are often unhealthy and uninspiring. In going back to more 'natural' and 'low-carbon' materials such as timber, earth and materials based on crop wastes, we might rediscover building techniques and designs that remind us of more congenial ways to live, i.e. with more stable temperatures, more natural ventilation and more connection to the outside environment. On the other hand, we build upwards not just because it's fun, but because it means less incursion onto land which might otherwise be wanted for agriculture, wildlife, or just breathing space. Perhaps one day we can have the best of both worlds.

Paper or electronic?

If more wood was to be used for building, would we still want to use it for paper? And even if there is enough to go around, and we could ensure that all timber and paper pulp production is from certified, sustainable sources, wouldn't we still want to 'design out' wasted paper? At one level this could mean just not using paper if we can help it, especially if it can't be recycled – everything from abandoning the use of toilet paper in favour of washing (as discussed by Rose George) to only having 'fastfood' that can be eaten in the hand and requires no packaging (as trialled by some elightened events organisers). There would still be books, since I cannot imagine a world without books on paper, but they are 'print on demand' at a price that reflects their environmental cost, so that none have to go through the ignominious and wasteful process of re-pulping, and so that second-hand books have high resale value. The same could be true for magazines and even newspapers, leaving things that people are happy to scan (and store or delete) to the electronic media. In the second quarter of 2010, sales of e-books on Amazon outstripped hardbacked books for the first time.

But is electronic really better than paper? It would depend on what else has changed. According to Mike Berners-Lee's figures, an email takes a sixtieth of the carbon used to produce and post a letter – but, as he points out, that is only a saving if we send the same amount of emails as we used to send letters, and how many of us do that? Another calculation shows that one email is equivalent to two sheets of virgin toilet paper, and that a single websearch is probably about the same. None of this takes into account the energy needed to make the computer. A more telling figure is that an average book accounts for 1kg of carbon, whereas Berners-Lee guestimates an electronic book reader at around 50kg. That's the 'embodied' energy to make it, without the electricity needed to charge it or the much larger power needed for the servers that provide the Internet. Quite a long payback period, if your concern is to save the energy used in paper. There is then the business of recovering the gadgets and reusing them or recovering their materials. As if to underline the point that electronic media are highly energy-intensive, the site of a closed paper mill in Finland has been reoccupied by a big Google data centre, at least in part because of the large supply of power already serving the site.

In future, if all energy is renewable, if water can be more efficiently recirculated in manufacturing processes, if the plastics are plant-waste-derived and the metals sourced responsibly, and electronic devices are designed to last and be only minimally upgraded, the footprint will be lower. But all this goes to illustrate that, just like paper, electronics can be done well or done badly, and the relative merits of both will depend on the management of each.

There is another consideration. The oldest book we have was made in the ninth century, and the Babylonians left us observations of the stars baked onto 2,500 year-old clay tablets. According to an article in the *New Scientist*, even the best forms of digital storage (including CDs, tapes and flash drives) can't be expected to last more than a hundred years. In an increasingly information-rich society, if we want to be sure to preserve vital aspects of our culture for future generations, we will have to either constantly refresh our digital archives, or perhaps commit it all to paper.

14

Everything designed for recovery

The second principle is of keeping resources in productive use. Evie and Ed's world is not one of zero waste in the literal sense. Zero waste, in the sense of nothing discarded, is not possible. Entropy, the unstoppable movement of stuff from order to chaos, ensures that whatever we use, for whatever purpose, will at some point need to be discarded or moved on. Creating a more cyclical economy is an uphill battle against both entropy and current economics. Zero waste represents an aspiration to let as little stuff out of the economic system as possible, and while a great deal of creative thinking has been done on the ways we might go about this, there are woefully few mainstream examples to point to.

Cradle to cradle

For a long time, using resources better has been characterised as 'resource efficiency'. This is not language familiar to most consumers, but when used by academics and industrialists it tends to mean getting more output for a given input ('getting more from less'). More recently it has also been used to mean reuse and recycling, so avoiding the final disposal of products and materials before their useful lives are really over, on the grounds that not to do so is inefficient. And indeed, as we saw earlier, we can point to some real improvements in the efficiency of the way we use resources, and greater levels of recycling. Measuring it as the amount of primary materials needed to generate a given level of economic wealth (also called material and energy intensity), we find we are doing much better now than in the immediate postwar years. But that says nothing for the *absolute* amounts of resources taken, and the amounts of wastes discarded, which have gone up hugely, with associated huge environmental impact. We are still doing the wrong kinds of things, but we are doing them more efficiently.

Michael Braungart and William McDonough, in their seminal book *Cradle to Cradle*, make this point very forcefully. Eco-efficiency is not the goal. 'If a system is wrong, and you make it more efficient, you are simply making it efficiently wrong.' The goal should be not to be less bad, but to be good. Every product should be a positive for the environment – taken from controlled extraction and then kept in circulation for as long as possible (if non-renewable), or returned to the land if a renewable resource. The idea is to establish two cycles – what Braungart and McDonough call a 'technical' cycle of non-renewable materials such as metals and aggregates, and a 'biological cycle' of anything that is harvested from the land. If these cycles are respected, they argue, absolute quantities of stuff are less relevant. Nature works like this – lots of stuff around, lots 'discarded' by plants and animals, but all of it broken down by nature's cycles and returned to use without compromising those cycles in any way. *Cradle to Cradle* asserts that there is a greater weight of ants on the planet than humans, but we don't notice them because their activities are part of these natural cycles.

So cradle-to-cradle (C2C) design means making sure that products can easily be broken down to their constituent materials – examples of C2C-certified products include floor coverings made from a single recyclable polymer, a type of Japanese lime plaster called 'shikkui' that is both recyclable and biodegradable, and benches made from single pieces of steel. It also means avoiding use of any additives that might hinder recycling, particularly if they are toxic to people or the environment – these are first in line to be 'designed out'.

This is a beguiling ideal. Braungart and McDonough's world is one of abundance, creativity, beauty and 'nutrients', rather than one of frugality, self-denial and narrowly construed 'efficiency'. Braungart is a chemist and McDonough an architect, so what we have is an interesting mixture of materials insight and design aspiration, but first and foremost this is a design book. It has some examples of cradle-to-cradle design, and these are, like the thesis, inspirational. The trouble is that they are just that – examples. Seven years after the book's publication, there are still relatively few cradle-to-cradle companies. The key limitation is that of the market. Braungart and McDonough think that companies will embrace cradle-to-cradle for no other reason than that it's a good idea. Which it is, but it's not without cost. If others are *not* doing it, those that are will be at a competitive disadvantage, however much they might want to change.

The cradle-to-cradle concept is important because it urges us to explore more fully how close we can get to the functioning of natural systems. It is also invaluable for its plausible vision of a positive, beautiful future, a feat that few environmental commentators have matched. But at the end of the day, it is likely to take a lot more than the sporadic attention and motivation of the private sector to move us to a comprehensively cradle-to-cradle world, which is why I explore the case for greater government intervention in Part 6.

Biomimicry

Very close to cradle to cradle in its principles is the emerging art and science of 'biomimicry'. This does what it suggests, taking inspiration from natural systems to design ways of living that are more in tune with the environment. Janine Benyus, the pioneering US exponent of biomimcry, makes the point that in nature, materials are expensive (their production takes precious energy) and 'design' is cheap, in the sense that the organisms around today have had the benefit of a 3.8 billion-year research and development phase called evolution. Michael Pawlyn, who worked with the team of architects who designed the Eden project's giant greenhouse 'biomes', is one of the principal UK exponents of these ideas. He makes the point that the effectiveness of some of nature's structures is hard to better. Our bones are constructed to use materials exactly where the stresses will fall and nowhere else. Birds' skulls have evolved for both strength and lightness by using a lattice structure similar to those used in the supporting structures for buildings. Leaves in the dim understoreys of the rainforest achieve large surface areas for light capture without being too thick by using complicated folds and ribs.

The same is true of nature's systems – they deal seamlessly with 'wastes', in cradle-to-cradle fashion, by deploying a series of natural circulatory processes, or 'closed loops'. The ability of plants and animals to biodegrade, or break down into their constituent molecules, is key to this circulation, but nature has also evolved some sophisticated ways of ensuring that materials do the jobs they need to do before being broken down. Janine Benyus gives the example of the protein threads that mussels use to attach themselves to rocks. The mussel manufactures the thread, and the glue that attaches it, all within minutes. It then coats the tether in a sealant that keeps out microbes, delaying biodegradation until after the mussel has moved on. Scientists are now avidly researching and hoping to imitate this ability to produce materials that can return to natural cycles, but at a time of our choosing.

Bioplastics – what are the advantages?

The mussels and their clever tethers take us neatly on to bioplastics. Plastics exemplify the 'throwaway society'. Hardly a week goes by without a press story about the banning or taxing of plastic bags somewhere in the world, but while sending an important signal, these measures are the tip of a very large iceberg.

In Evie and Ed's world, dependence on oil has come to an end, with energy coming from a wide variety of clean sources. Plastic materials are now made from plant products, most of them crop wastes, or even food waste, rather than growing the feedstocks specially. As far as possible they are reusable and recyclable many times, and when this is no longer possible, they are properly biodegradable and so can be digested along with other organic wastes to recover energy via the gas. The residue can be returned to the land.

Bio-based plastics are made by using plant sources of polymer rather than oil, and by employing some clever chemistry. There are easy ways to get bioplastics and hard ways. The easy way is to use carbohydrates (starches and sugars) from major food crops such as maize, wheat, sugar beet or potatoes. These substances are the plants' energy stores – they are also what we eat to give us energy. We wouldn't want to use these sources if it meant going into competition with using them for food, although we already use nearly half of starch produced in agriculture for industrial uses without affecting food supply. The hard way is to convert cellulose, a natural polymer (long-chain molecule) of sugar, which plants use to strengthen cell walls and which we can only partially digest – we call it dietary fibre. The even harder way is to break down woody materials and convert their lignin – lignin is what gives trees and other woody plants their superior strength. The last two processes are more energy-hungry than using starch, but if driven by renewable energy in very efficient 'biorefineries' they have a much better carbon profile than oil-based plastics.

This variety of methods means both good news and bad news. The good news is that it can be relatively cheap and easy to get plastics from plants, but the bad news is that if using the easiest route, that may mean direct competition with food and animal feed crops. Conversely, it may require more energy, water and chemicals to process cellulose and lignin, but these have greater potential to be derived from plant wastes.

Bio-based plastics are often biodegradable, though not always. Some plant-based plastics are being developed because they use less energy to produce than oil-based plastics, and this is a bigger motivation than the fact that they come from a 'renewable' feedstock, or that they can break down in the environment. Confusingly, there are also some oil-based plastics than can degrade (as in break down into tiny pieces), and even completely biodegrade (i.e. break down into their constituent molecules and return to nature's cycles). But the ideal 'plastic' will come from plants, with minimal use of energy, water and additives to process it, will be completely recyclable, and will also biodegrade completely.

Interest in bio-based plastics and, as a subset of them, biodegradeable plastics, has burgeoned with increases in the price of oil, and concern about the greenhouse-gas implications of using fossil feedstock. Production of these novel materials was less than 1% of global plastics production in 2007, though the market in Europe is increasing rapidly. Even so, the authors of a major review in 2009 analysing the prospects for bioplastics were not optimistic about take-up, expecting maximum market share at only 4% of all plastics by 2020. More recent estimates put it at 20%, even though on a theoretical basis 90% of plastic polymers could be replaced by the new materials.

What are the limitations? Primarily cost. These processes are in their infancy compared to conventional plastics, and they have not yet benefited from the same experience or scale of production. Some are more complex than oil-based polymer production, and the technical glitches have yet to be ironed out. They also, of course, have to compete for investment with conventional plastics, and that depends to a large extent on the price of oil. When the oil price was at its highest in 2009, prospects for bioplastics started looking up, and anything above $50 a barrel makes them competitive.

Work by Green Alliance has highlighted another major barrier to take-up, at least in the UK. Green Alliance's 'Closing the Loop' and 'Designing out Waste' projects worked with companies spanning the whole supply chain (manufacturers, retailers, waste companies and trade associations) to examine how bio-based, biodegradable (and therefore hopefully compostable) plastics might best be deployed. This work was started because supermarkets had pounced on bioplastics for things such as the windows in sandwich wrappers and potato bags, and were loudly labelling them as environmentally-friendly. What they had not really considered was what consumers were supposed to do with them. Although they have other benefits in terms of taking less energy to produce, it would be a good idea for materials that are compostable to actually get composted, not least because they would release methane if left to degrade in landfills. However, in 2007, only around a third of households claimed to compost at home, and probably fewer than this were doing it effectively.

Worse, it wasn't clear which of the materials on the market could be composted at home, and which required the higher temperatures of municipal composting. There wasn't even an agreed labelling scheme, indicating which route the new plastics should follow (assuming that home or municipal composting was available), and it took another three years to develop such labels. But that left a disproportionate burden on householders to put the right materials in the right receptacles – materials that were often indistinguishable at first glance – if no advanced sorting was available at the recycling plant to make sure that they went the right route.

The people who really worried about this were the people at the very end of the chain, the people operating the composting plants where the local authorities were sending their kitchen or garden waste. They were distinctly cool about having to deal with bio-based plastics – after all, how would they tell whether they were compostable plastic or conventional plastic when they came in? If householders were lazy or mistaken about which they put where, the processors could end up with both types with no way of knowing. The only way to deal with that would be expensive sieving of the material either before or after composting. These concerns in turn made local authorities also cool towards bioplastics, and retailer interest began to wane.

Hannah Hislop at Green Alliance worked long and hard to think through how this impasse might be broken. Discussions suggested that what was needed was clear guidance about how the new materials should be used, in order to develop a logic that was compelling for consumers. Compostable plastics for compostable products seemed a good place to start – concentrating the use of bioplastics on fresh produce, flowers and even sweet wrappings, and labelling and promoting them thoroughly. Although this logic was widely applauded, so far it has had patchy results. At an update meeting in 2009, the innovation of the bioplastics industry was demonstrated by products such as gold foil tea-bag wrappings (biodegradable plastic with an extremely thin layer of metal), which although ingenious and technically meeting the relevant standards, and even appropriately labelled, looked about as compostable as a kitchen knife. Without widespread publicity to inform about the composting route, these products risked confusing consumers.

What all this demonstrates once again is the imperfection of the marketplace. Everyone claims to want more 'environmentally friendly' materials, to move away from oil, and to have more recycling and composting. Yet materials that could deliver all these benefits are hamstrung by lack of a clear industrial or political vision, or a route map to widespread deployment.

This situation can only be changed by standards for products that emphasise the required qualities. If product specifications for certain applications (such as all fresh-produce wrappers) stipulate that they must be compostable (including being home compostable), then they will be, and everyone will know that they are. That includes consumers, so they are less likely to make mistakes about what they do with them, and also the processing facilities, who can be sure of a certain level of these plastics in the material they take, and will adapt to treat them, including dealing with small but expected levels of contamination.

Using plants for something as ubiquitous as plastics would entail the use of a great deal more plant material (and much better utilisation of waste plant material) than ever before. The question is whether there would be enough land to go round.

The debate over biofuels that raged in 2008 and 2009 has started to give us insights into these issues. Biofuels are plant-derived substitutes for petrol and diesel, considered desirable because although burning them releases carbon dioxide, it is CO_2 that has been taken up during the life of the plant and so from the existing store in the atmosphere, rather than fossil carbon which is continually topping up that store. Some commentators are very concerned that any expansion of biofuels, or biomaterials such as plant-derived plastics, will exacerbate existing pressures on habitats, by increasing the amount of land that has to be devoted to agriculture.

The worldwide food price rises of 2007/2008 were blamed in part on suddenly diverting land towards biofuel production, reducing the amount of grain available for food, particularly corn, and thus driving up prices. To make matters worse, there seemed to be evidence that growing some biofuels might actually lead to greater carbon emissions, once indirect land use changes such as clearing rainforests to grow the crops had been taken into account, to say nothing of the loss of biodiversity that also results.

In the UK, however, a major study on biofuels, the Gallagher Review, concluded that available land globally could support agriculture for food *and* biofuels, especially if currently idle or less productive existing agricultural land was utilised, and provided that maximum use was made of crop waste products. There was also the proviso that stringent sustainability standards should be in place, to ensure that biofeedstocks really do have carbon benefits, and do not encroach onto wild lands. The Gallagher Review also highlighted the possibility of producing biofuels with 'co-products', such as animal feed derived from the crop residue. Others have suggested 'co-cropping' – either growing a quick biofuel crop before a food crop is planted, or growing the two together in a mixed system. To further improve the carbon performance of biofuels, WWF has examined the concept of 'closed-loop biorefineries', employing anaerobic digestion and enzyme technologies, to maximise the amount of energy captured from the crop.

All of these principles could be applied to the production of bioplastics. So bio-based plastics could offer a way to secure materials from renewable resources that range from the very durable and recyclable, kept in circulation almost indefinitely, to the very lightweight and ephemeral, sent to digestion or composting very quickly. It is also possible that oil-based plastics could coexist with these new materials, but with the non-recyclable versions designed out. Both types of materials could subscribe to cradle-to-cradle principles if done well.

15

Nutrients cycled

Cradle to Cradle talks of all materials, renewable and non-renewable, degradable and non-degradable, as 'nutrients' in a future, much more positive, version of industrial society. More commonly, however, we think of nutrients as the substances that make things grow – chiefly carbon, nitrogen and phosphate. We interfere with the natural cycling of these nutrients by loading the environment with excess carbon from burning fossil fuels, as well as excess nitrogen and phosphate from the use of synthetic fertilisers. At the same time, we fail to return to productive use a large amount of potential nutrients from food and animal wastes, as well as our own sewage. To solve this means redesigning systems on a grand scale – we are not talking about something as relatively straightforward as changing the plastic used to make a drinks bottle.

Wasted food

I hadn't meant to write much about food and food waste since there has been a wealth of recent literature, and the complexities of global agricultural production and economics merit whole books in themselves. However, I could hardly paint the picture of Evie and Ed's world without including food. My thoughts about what would be 'good' food systems in the future centre on reducing meat consumption, because producing meat takes so much more resource than producing grain or vegetables, especially if the animals are fed with grain rather than through grazing. There is also the problem that livestock produce methane, a much more powerful greenhouse gas than carbon dioxide – burping cows really are warming the planet. Equally important is a moral prohibition on wasting food. Another key factor is the ability to deal with unavoidable food waste at household level, to produce both energy and nutrients for the soil. Behind the scenes, losses of food in transit (which is often at shocking levels, particularly in warm countries) should be avoided by investment in proper storage.

The issues around food waste are perhaps the most dismaying. Food waste easily qualifies as one of the 'worst' wastes in terms of the amount of energy, water and material resources that go into its production. It should also be the most avoidable. Food is for eating, and yet, annually, here in the UK we fail to eat some 25% of what we bring into our homes – WRAP estimates the amount at 8.3 million tonnes. That is without the losses through the supply chain: at the farm, in transit (often from far-flung parts of the world), in manufacturing, and from shops and restaurants. For the US, the total has been estimated as at least 40%. Globally, waste is hard to estimate because few countries count as well as the UK does, but a recent report for the UK Government quoted estimates between 10% and 50%, with much higher figures for some crops. Writer and campaigner Tristram Stuart argues that avoiding these losses would provide everyone on the planet with all the calories they need without further expansion of agriculture.

Part of the reason for the waste is, as ever, economics. Agricultural subsidies encourage overproduction in some parts of the world, and impoverish farmers in others because the subsidies keep world prices depressed. Thanks to globalised production and cheap labour, consumers in the developed world spend a lower proportion of their income on food now than they did twenty or thirty years ago, so they can 'afford' the waste. But globalisation itself comes with a hidden cost – now that the UK imports 40% of its food, it has also become responsible for the waste in that part of the supply chain. Post-harvest losses in some parts of the world are enormous – sometimes as high as 50% – for lack of investment in simple technologies such as grain silos and cold stores. The increasing move from rural areas to cities in many countries makes this worse, as food supply chains get longer and more complex. This is before any food waste is incurred in the UK from the manufacturing and retailing parts of the chain – 7 million tonnes a year of it to go alongside the more than 8 million tonnes that consumers throw away. It is often cheaper for companies producing and selling food to waste it than to find alternative uses for surpluses. Also, as people get more affluent, their diets diversify away from the starchy staples to more fresh fruit and vegetables, which are more perishable; meat and fish, which are more resource-intensive to produce; and dairy products, which have a high carbon and water footprint. All this means that wasted food has an increasingly high environmental cost. The economic system does not factor in the environmental damage of food production any more than it does for other commodities, so prices give us a false illusion of plenty. Food is cheap and waste is even cheaper. The waste of fish is particularly scandalous – over-exploitation is destroying fisheries worldwide, and yet some 25% of sea-caught fish never makes it onto a plate. So overall, as much as 40–50% of food is wasted before it even gets to us, only to have us waste a further 15%–20% because we overbuy, get portion sizes wrong, or simply don't notice it going off.

The Western illusion of limitless plenty is in front of us every time we visit a supermarket, any time of day or night, and the shelves are full. When I was a child, if you visited the shops (including the local, tiny, emergent Tesco) any later than 3 p.m. in the afternoon they would very likely have run out of bread and other staples, and there were no more than two or three choices for any given product. This is no longer allowed in supermarket-land. Retailers believe that empty shelves or restricted ranges will drive consumers into the arms of competitors who can maintain the artifice of constant availability, be it strawberries in the depths of winter, or fresh bread in the middle of the night. We shouldn't be surprised if there is waste at all stages – from the supermarkets, from their suppliers who order in too much for fear of not giving supermarkets what they want, and from our homes when we're tempted to buy more than we need. I once dared to suggest at a dinner with senior executives from one of the major supermarkets that the best thing the store could do for the environment was to run out of food from time to time. This was met with a frosty silence.

How would these pressures be addressed in Evie and Ed's world? One answer is that food would cost more relative to other commodities, thanks to a gradual process of factoring in the environmental costs, at least in the rich world where it can be afforded. Also, it would be almost impossible to dispose of wasted food other than at high cost, since landfill would be a thing of the past and access to industrial digestion plants restricted to the most unavoidable organic wastes. Food waste from households could be collected and sent for processing into animal feed, but householders would have to pay for the collection service. So the whizzy household technology to process scraps and sewage into gas and soil conditioner would be one response to these pressures (not the same as sink waste disposal units, they flush food waste into the sewer system, which would be severely overloaded if everyone did it). The technology is important, but only a small part of the story. The overarching achievement would be that of governments globally making people understand, and if necessary pay for, the enormous environmental cost of food. They would then be more inclined to understand its value.

In the UK in coming years, there is a risk that focusing on a 'technological fix' will distract from this fundamental need to persuade people and businesses to stop wasting food. Of all waste-treatment technologies, anaerobic digestion (AD) has been singled out for the most encouragement and financial incentives. AD is a means of processing food, crop waste and manure, and sometimes specially grown crops, to produce gas and a nutrient-rich liquid called 'digestate'. To picture an AD plant, think of a large, liquid compost heap sealed inside a giant tin can. Over a period of weeks, the waste is broken down by a population of microbes in an oxygen-free (anaerobic) environment, to produce biogas – a mixture of carbon dioxide and methane which can be used as a fuel to generate energy. If the stuff going in has been sufficiently well controlled to keep out contaminants that might compromise growing food, the 'digestate' is a good fertiliser for farmland. One state-of-the art facility I visited in the Netherlands produced a range of saleable composts, including some with in-built pest-control.

The availability of this technology, actually very old but only recently developed as an industrial-scale solution for wastes in the Netherlands and Germany, has been seized on by the UK and other European countries looking to solve the problem of 'biowastes'. In political terms, biowastes are not a problem because of the waste of food, or because the waste of matter that could be an agricultural nutrient is felt to be wrong, but because EU legislation is gradually ratcheting down the amount that can be buried in landfill. In landfill the same thing happens to biodegradable waste as it does in an AD plant, but in a less predictable and controlled way. Even though some landfills have pipework and engines to make use of the methane produced, much of it leaks to the atmosphere where it has twenty-five times the global warming effect of carbon dioxide. So EU legislation limits biodegradable waste to landfill (although mainly from household waste, and not from all businesses) in an effort to reduce its contribution to climate change.

AD ticks lots of boxes in the UK. It helps to meet the EU landfill directive, and it also provides a 'renewable' source of energy (i.e. derived from a biological, renewable resource rather than fossil fuels), albeit at a low level for the foreseeable future. Although the National Grid company has said that it might be possible to meet as much as 50% of UK domestic gas requirements from biogas, this is based on some of the most optimistic of their scenarios.

AD can also rightly be described as a 'closed loop' for nutrients – resources that come from the land go back to the land. But let's be under no illusion that this can ever justify the appalling scale of avoidable food waste. WRAP has estimated that for every tonne of food waste saved, 4.3 tonnes of CO_2 equivalent emissions are prevented. For every tonne of food waste going into AD it is just a tonne of CO_2 equivalent emissions. On this measure, it is therefore more than four times better to prevent the food from being wasted in the first place.

Taking to heart the true scale of food waste and tackling it could put enough extra nourishment into the global food system to avoid needing more land for agricultural production, indeed it could even free some up. It has to be one of our foremost goals.

What we should do with our poo

I have visualised for Evie and Ed a household device that digests not just food scraps but also their own waste. Why is that a good idea? For two main reasons: human excrement is a source of precious nutrients and we shouldn't waste it; and using clean water to flush toilets is a waste of both water and the energy used to clean it to drinking-water standards and pump it to our homes.

In *The Big Necessity*, Rose George explores sanitation (to be more frank, toilet habits) around the world, in search of the answers to what is still the biggest killer of children – disease caused by living among human waste. Globally, four in ten people don't have access to sanitation of any kind, not even a hole in the ground, meaning that excrement is left anywhere and everywhere, spreading disease and death. At the other end of the scale, the Japanese have invested millions of yen developing the world's most sophisticated toilets, computer-controlled, with inbuilt washing and drying of essential areas, heated seats and self-cleansing mechanisms. George has a very entertaining account of the research effort that went into finding exactly the right angle for the water-spray nozzles to ensure complete personal cleanliness in these most advanced of loos.

We take the luxury of sanitation largely for granted, but until the nineteenth century, towns and cities in the developed world would have been as rank as anywhere else, with overflowing cesspits and chamber pots emptied onto streets (the word 'loo' may have come from the '*gardez l'eau*' shouted to warn passers-by). The flush toilet is an invention of the West. The first models were developed in Britain (opinion seems to be divided as to exactly who was responsible, but it wasn't Thomas Crapper, he just made them, and the word 'crap' comes from an old Dutch word meaning 'rubbish'). They gained popularity as sewage systems developed in big cities in the nineteenth century, since it was then easy to flush the stuff away to somewhere out of sight and smell. However, the sewers weren't able to magic it away, they simply put it in the river further away from human settlements. Eventually, as populations grew, this too became a problem, especially where rivers were also a source of drinking water. Sewage treatment started in the mid-nineteenth century, so that the effluent that finally went into the rivers or sea had been at least partially cleansed. This was driven much further in the twentieth century by the fact that the EU started to raise the standards for the quality and appearance of drinking water.

So by the time the water companies were privatised in the UK in 1989, we had excellent drinking water, but also a system where water cleaned to very high standards was being flushed into sewers and mixed with all kinds of filth (because industries as well as households discharge liquid waste to sewers), which then had to be taken out at sewage works at great expense. All these water-cleansing facilities take energy to operate the many filters and pumps. It has been estimated that UK sewage treatment is responsible for 5 million tonnes of CO_2 emissions, equivalent to burning over a million tonnes of coal.

Having cleaned up the water that comes with the sewage, there remains a 'sludge' of microbiological residues from the treatment process, plus any pollutants that the treatment can't remove. This has to go somewhere. We used to dump most of it at sea, but this was outlawed in the mid-1990s, again by the EU. Like any manure, it is a potentially useful source of soil conditioner and nutrients for agriculture, with less environmental downsides than manufacturing artificial fertiliser as we mainly do now.

As historian David Edgerton explains, switching from largely locally available human and animal wastes to using fertiliser from deposits of bird poo, or 'guano', boosted agricultural production hugely in the late eighteenth and nineteenth centuries. The stuff was mined extensively in the Caribbean, Pacific Islands and Peru, and served everyone very well until it ran out. Then the process of hydrogenation, invented in Germany between the wars, enabled the production of nitrate fertiliser from oil. This was arguably a more important, or an equally important, innovation as the development of the hybrid crop varieties that are credited with the 'green revolution' in agriculture, because without the fertiliser they could not have reached their full potential. But artificial nitrate takes energy to produce and is made using a non-renewable resource. When human waste is returned to the land it saves us some of that energy. Sludge also has good quantities of phosphate to replace mined sources, particularly if the right 'phosphate stripping' techniques are employed at the sewage works. In fact, according to some estimates, there is enough agricultural nutrient available from human waste, municipal food and garden waste, and farmyard manure to replace all use of manufactured fertilisers.

However, putting sewage sludge back on the land is limited by needing to take care about pollutants that may have gone down the sewer too – whether household or industrial chemicals, or just unacceptable concentrations of the metals that we excrete from our bodies. In addition, some retailers are sensitive about letting people know that crops are grown on human waste, even though it has been treated, as if it is somehow more distasteful or dangerous than animal wastes. And, although research has shown that health risks are minimal, it can smell bad, although no more so than other animal manures. Adding up these concerns, we end up with a situation where only two-thirds of human waste nutrients are returned to farmland. The rest is either incinerated (and the ash goes to landfill) or used for 'land restoration' – it is useful mixed with mining waste, for instance, to create a new growing medium on reclaimed sites. But recent research has shown that both farmers and consumers are quite comfortable with the idea of food being grown on human, animal and composted household wastes, provided the right quality controls are in place.

So the flushing toilet and associated sewage system

solve sanitation problems by using up one precious resource to remove another potentially useful resource, and we don't at present do very well in terms of 'closing the loop' and getting the nutrients back into the growing system. This doesn't seem logical, and when we realise that some countries are considering rolling out Western-style sanitation to already water-stressed areas, it seems like lunacy.

Rose George chronicles the arguments between those in the sanitation industries who think that only flushing will properly address health concerns, and those who think that there may be more environmentally friendly technologies. 'Ecosan' (ecological sanitation) or 'composting' toilets are being promoted for use in the developing world, where water supplies are not secure enough to establish the kind of sewage infrastructure we have here, or there simply isn't the money to build it. They are also growing in popularity in parts of Europe, including Germany and Sweden, where in two municipalities it is now the law that new toilets must be of this kind. We encountered one while on holiday on a Swedish island, much to the bemusement of my two boys, because they require males to pee sitting down. There are two compartments, so that urine is separated from faeces. The faeces can then dry and compost with relatively little odour, and the urine can be collected to use as fertiliser. It is surprisingly valuable stuff – it has been suggested that, under the right conditions, the nutrients in one person's urine are sufficient to grow 50–100% of their own food requirement. The composted dry matter can also be used, and indeed at the Centre for Alternative Technology in Wales, all the products of the visitor toilets go on the fruit trees.

In parts of China, this has been taken a step further. Human waste is mixed with pig manure and digested in miniature, household AD plants, which then provide useful 'biogas' for cooking and for lamps as well as yielding fertiliser at the end of the process. In rural areas where there is no electricity, and women habitually spend hours collecting firewood to cook on, and then more hours breathing the smoke, this is a big step forward. The technology is not foolproof and can be hazardous, so commitment and careful training are required. However, I feel it's a sufficiently encouraging story to foresee in Evie and Ed's world an updated version of this sanitation/energy/water-saving device to deal with their poo, eat up their few food scraps, and fertilise their flowers and veg. In their small way, they are helping to close the nutrient loop.

There is very long-term benefit suggested here. If we can deal with human waste safely 'on-site', as it were, we will save a great deal of water presently used to flush it away. In addition, manufacturers are developing waterless (or nearly waterless) washing machines, and dishwashers may be next. If the water used in the home can be kept to a minimum, stored rainwater might provide the majority of the supply. If it can then be kept free of pollutants (by using fewer household chemicals), the waste water from houses can be used to water gardens – this is already strongly urged by those trying to relieve pressure on fresh water sources. So what arrives in the tap is almost all for drinking, and what goes down the drain is kept to a minimum. Eventually, we might be able to do most water treatment through smaller-scale community-level plants, cutting out a lot of the expensive and energy-hungry infrastructure that is modern sewage treatment, and looking after our water sources more effectively.

It is hard to see how this approach would work in towns and cities, although fans of ecosan claim it can be done. But someone has to produce the nutrients we need for farmland to replace artificial fertilisers, so we'll also need versions of the composting toilet that allow collection of the products for safe use in agriculture. The future echoes the past – in London before flushing toilets and the sewage system, the 'night-soil' collectors performed a vital role in transporting waste to the farmland surrounding the city. Evie and Ed's world might feature a much more high-tech, odourless version of this practice.

Nappy sense

As a mother and an environmentalist, having talked about paper, plastic and poo, I can't resist a short digression on the subject of nappies. I'm going to treat them as 'nutrients', or at least hope that they can be in future.

Evie and Ed will be able to have the choice of using a form of disposable nappy for their children (because these are designed to fit in with their household food scraps/poo digestion technology) or using washable ones. They will, however, be sure to have their children out of nappies as soon as humanly possible – however 'green' the nappy, it is still a use of resources.

According to William Rathje, who devotes a whole chapter of his book *Rubbish!* to the subject, the disposable nappy was invented by Johnson & Johnson after the Second World War, but really took off when Pampers were patented in 1961 by Procter & Gamble. The main ingredients are paper pulp, a coating of plastic, and, in between, a magic gel, a manufactured polymer called sodium acrylate. This has the very useful property of holding fifty-seven times its weight in water.

Use of disposable nappies is increasing rapidly world-wide. It is claimed that the average UK baby uses over 4,500 before being potty-trained, and 3 billion are thrown-away every year. It is likely that children stay in nappies longer because of the increased convenience and comfort compared with the old-fashioned, soggy terry variety. My mother-in-law, who was a nurse and then a health visitor, was sure that disposable nappies aided children's early development because of the greater freedom of movement they afforded. When my elder son was born in 1997, we experimented briefly with reusable nappies, but since we had hardly any drying space and no access to a pickup service, we lasted about a fortnight. My mother-in-law clearly thought we were barmy to even try. Evidence as to which is 'greener' has been hotly disputed through various 'life-cycle' studies, none of which is necessarily comparing like with like or taking account of the reality of life for harassed parents. Fourteen years after we were agonising about this, nappy laundering services are more widespread and the design of reusables has improved immeasurably, so uptake is growing in the UK.

The chief objection to disposables by fans of the reusable variety is that they are filling up our landfills. That's true, but arguably no more than many other paper and plastic products. Nappies make up around 3% of the household bin in the UK, compared to 23–25% that is other paper, and 8–10% that is other plastics. If the argument is that the wee and poo is generating methane, it is a small contribution beside all the food and garden waste sent to landfill – totalling as much as 38%. Even pet faeces might make a comparable contribution – in San Francisco these have been estimated to be at almost the same level as disposable nappies in household waste. But they are the largest non-recyclable fraction of our bins, and if we are going to aim for Evie and Ed's world, all these materials and products will have to subscribe to the cradle-to-cradle principles we've been discussing. So what counts is how we take nappies, whether disposable or reusable, down that road.

One option is certainly digestion. This would be possible if the whole nappy were made from plant-based products – so paper pulp (preferably from crop waste rather than trees, as discussed in the section on paper), plant-based plastic and a gel that degrades. It is not clear whether the sodium acrylate used at the moment does fully degrade, but work is in train to develop alternatives

– so far the most promising is the wonderful-sounding 'thermal polyaspartate' which does degrade, and has similar liquid-retentive properties. The very enterprising AD pioneer Julian Maikelm in Cornwall has experimented with digesting soiled nappies on a small scale by putting them through a garden shredder and then adding them to a farm-scale anaerobic digestion plant. He has shown that even with current nappies it can be done (although the non-degradable plastic has to be sieved out, and cleaning the shredder is a bit tricky). But it may only take minor design improvements for it to work in the household-scale digestion technology used by Evie and Ed. And reusables would work on the same principle – liners that can go in the digester and cloth (preferably not cotton but alternative, lower impact crops) that can be washed. Fertiliser from the digestion goes on the garden, helping to close the nutrient loop. When they're older, the babies will be proud of this destination for their offerings.

16

All energy renewable

We need energy to heat and power our homes, and to give us transport. In addition, everything we consume and then send away for reprocessing, be it meat, mobiles, sewage, washing machines or waste to landfill, requires energy input at some stage of its life cycle. And we need energy to beat entropy – to gather everything up and return it to being useful by powering all the systems we need for recovery and recycling of materials.

How much energy can we have?

There are two aspects of energy and stuff that we need to worry about. The first is the supply – where do we get energy from, and what are the environmental consequences of the various sources? The second, just as important, is about demand – how much do we use? It would be tempting to think that if we crack the first and come up with clean, green, cheap energy sources, we needn't worry about demand. This is over-optimistic. Any kind of energy infrastructure will have a material cost which means that the more we can conserve energy, the easier will be the job of keeping our activities within planetary boundaries.

So making the energy equation add up without using fossil fuels, as has been managed in Evie and Ed's world, involves finding new sources of supply but also ways of limiting demand. This may mean more than simply striving for efficiency (which implies doing the same things, but doing them with less energy or stuff) because we may make up for efficiency gains by using the opportunity to do more of something else. As described in Chapter 8, this is called the 'Jevons paradox' or 'rebound effect' and has been the story of industrialised nations since the Second World War. Resources are being used more 'efficiently, i.e. generating more wealth for fewer inputs, but this has simply generated the money and space to accelerate industrial activity even more, leading to greater use of resources overall. So genuinely conserving energy, or indeed anything, might mean setting a 'cap' for how much can be used in aggregate. This is a fairly radical line of thinking for most politicians – it implies 'rationing' resources, at an international, national and even personal level.

Where will energy come from?

There are really only four sources of energy: the sun, the moon (which drives the tides), the heat in the earth's core and the decay of nuclear elements. As Oliver Morton points out in his delightful book *Eating the Sun*, fossil fuels are not sources of energy directly, they are ways of storing energy – the sun's energy. It was absorbed by plants and animals and laid down a long while ago, by means of that extraordinary mechanism that turns green shoots and sunlight into carbon – photosynthesis. And although the environmental agenda is presently intensely preoccupied with how to move away from fossil sources, globally we get more carbon-based energy from burning more recent products of photosynthesis, in the form of wood and other crops, or crop wastes. This more recent plant-based energy (known as 'biomass' energy) is considered a better bet than fossil fuels from the point of view of global warming, because it has recently absorbed as much carbon dioxide to grow as will be emitted by burning it. This means it counts as a clean, 'renewable' source of energy. But it still needs to be used carefully, as the controversies around some biofuels have shown.

Why is this? As we've already considered, renewable crops such as wood, whether for fuel or for paper, are only sustainable if properly managed, so good forestry is the first criterion. It is also crucial that improving the global contribution of biomass energy need not compete with other uses of land, such as food. But biomass energy can come from a whole range of crops and crop wastes, preferably after they've already had another use, such as being made into paper or plant-based plastic. In addition, a great deal of wood gathered for fuel is used inefficiently, or in ways that cause local pollution. Cooking indoors on wood fires is a major source of lung and eye problems, particularly for women, in many less developed countries – a problem that can be solved by the introduction of relatively simple cooking-stove technology. Here in the UK, anyone who wants to burn wood to heat a home should be using a wood-pellet boiler or a wood-burning stove rather than an open fire, since these get much more heat out of the wood.

Biomass is not the only option for harnessing the energy from the sun. As Morton points out, most of the 'renewable' energies we have developed so far are 'solar', whether it is tubes warming water, electricity generated by photovoltaic cells, wind power (the sun's energy makes the wind blow), or waves (the wind makes the waves). Only tidal (driven by the moon) and geothermal energy (tapping the heat stored in the earth) are independent of the sun's radiation. It should not be beyond our collective wit to do much more with the star that rises every day on our horizon. 'The sun provides the earth with more energy in an hour than humanity uses in a year,' says Morton. He compares a photovoltaic cell with a leaf: the former is relatively efficient at converting the sun's energy to energy we can use, but it is expensive. The leaf is very inefficient but there are plenty of them. The trick is to find a technology somewhere between the photovoltaic cell and the leaf, something that can be incorporated into buildings (like solar tiles) or festooned in trees (keeping the leaves company without being too obstructive), so that solar energy can be harvested in small amounts by a great many units. Then the problem is using it immediately or storing it, because transporting it risks losing a lot on the way – storage of power from renewables is one of the biggest challenges on the scientists' to-do list.

Part of the problem is to do with the way we access energy. In the industrialised and industrialising nations we have evolved a dependence on fossil fuels burned directly for heat and power, but we have also developed an addiction to electricity. Converting sources such as oil, coal and gas to electricity is not very efficient anyway, but we do it on a large scale in power stations and then make it even less efficient by delivering it down cables which lose a lot of it on the way. It is, of course, possible to generate electricity on a local, immediately used basis through small generating engines, but we tend to prefer the convenience and silence of flicking a switch and having the current surge into our homes whenever bidden. We also have an addiction to portable electricity for our gadgets, so we have to manufacture, distribute and discard little parcels of metal and acid to feed that addiction, all of it at an energy cost in itself. In addition, we have proliferated uses for this inefficiently produced power because the price of it (here we go again) has been relatively cheap compared to people's incomes. And in the course of producing power, we have neglected to capture the wasted heat that accompanies it – everything from the heat that is let loose from power-station cooling towers to the heat that comes out of the back of your refrigerator.

Big kit, home kit and cool products

Green Alliance's work on energy has illuminated three main choices: 'big kit' (ways to generate electricity on a big scale and deliver it to the grid), 'micro-generation' (home-scale solar panels, wind turbines and other technologies) and 'cool products', re-designing all the appliances and gadgets we have around us to be as energy-efficient as possible, and, where feasible, generate their own energy.

The big kit has to be part of the picture. Most countries, certainly the industrialised countries, have a grid system for electricity and are not going to abandon it overnight, so in a growing number of countries the race is on to connect it to cleaner, renewable sources of energy such as wind turbines, wave machines, water-powered turbines in rivers (hydroelectric) and geo-thermal (heat from underground rocks). All of these have their pros and cons and there are many people more expert than me who can rehearse them. The hope is that as policies to limit greenhouse gases start to bite, particularly by putting a high price on carbon emissions, investment will flow to these options so that they gradually replace fossil sources of power. The speed of this is at present highly variable – rapid in countries such as Germany and Spain where they have thrown money at the problem in the shape of large subsidies; slower in the UK and US where it has been largely left to the market. The energy market in the UK has not, until very recently, seen sufficient certainty about government commitment to developing alternative energy sources to make heavy investments – it is gradually shifting in response to incentives, but not at the pace needed. Policymakers are also now looking seriously at how to create incentives to capture more heat.

So the new big kit is progressing, but not quickly in the UK. What about smaller kit? This is what I have envisaged for Evie and Ed. Green Alliance has been among the organisations arguing for the development of 'dispersed' or 'micro' generation – household-level devices that generate heat or power, which can also be connected into the grid if wanted. That way, surplus energy can be redistributed to others, or the grid can top up deficits in the house. Currently available micro-generation technologies include the kind of solar panels that heat up tubes filled with liquid to provide hot water, and the other kind (photovoltaic) that directly generate electricity in their clever silicon cells. Some panels now combine both technologies. There are wind turbines of the modest variety, and there are also 'ground-source heat pumps', the sinking of pipes several metres under the garden to tap the greater warmth of the soil down there and exploit the difference through a heat exchanger, which then provides low-level heat into underfloor heating. In addition, there are air-source heat pumps and domestic heat-storage devices that allow you to deploy heat in the home more flexibly. Air-source heat pumps can also be reversed to provide cool air to buildings on hot days. Developing technology includes cleverer kinds of boilers, such as ones that provide both heat and electricity. There are also boilers that burn biomass (which are basically updated versions of a wood-burning stove). None of these technologies is perfect, and all are still relatively expensive to install in terms of payback time for the householder, but then because the market has always been small, none has had the benefit of really serious investment and refine-ment. In turn, the market has remained small because they are not financially attractive and don't always work very well.

This seems like a real catch-22, but it may change with the long-awaited introduction in 2011 of 'feed-in tariffs'. These amount to a hefty bribe to householders to install these technologies and feed the electricity they generate into the grid. This electricity is 'bought' by the power company at a premium price, while power drawn from the grid remains at a lower price. This mechanism has been used in Germany and Spain with great success in terms of stimulating uptake of renewables, although some say that it has been a relatively expensive way of doing it. It is hoped that this mechanism will bring about an important shift – in the UK it is likely to bring the payback times for some technologies down from decades to less than ten years, boosting uptake. It might also boost some of the more innovative community-scale power options, such as the geothermal plant being proposed at the Eden Project in Cornwall, where the high temperature of the granite rock lying nearly four kilometres underground is exploited to create electricity. Even so, the most optimistic estimates of the contribution to UK energy needs put these *decentralised* renewables at only 15%, and the more pessimistic at only 1%, so in the short to medium term it is a complement to 'big kit' sources rather than a replacement.

Meanwhile, staggeringly little attention has been given to other ways of reducing or replacing demand for fossil fuels, especially through the way we design products. It is rapidly becoming clear that many electrical and electronic goods use far more power than they *need* to. The contribution of 'gadgets' to household energy consumption is not huge compared to heating, cooling and transport, but is more than the contribution from lighting, according to energy expert David MacKay. You might suppose that efficient use of energy would be a standard design criterion for any product, given the large consumer benefit in terms of taking money off bills. But even the very inefficient incandescent light bulbs have had to be phased out by regulation rather than as a consequence of manufacturer action or consumer demand. For other energy-using products, it is only recently that their performance has been addressed, through energy labels and energy-efficiency standards proposed by government, rather than by the manufacturers.

Both energy labels (to enable consumers to choose the more efficient products) and standards (which impose legally binding levels of energy efficiency) are initiatives of the European Union. Many have been contested vigorously by manufacturers, who see them as a threat to their position in the marketplace, rather than an opportunity to improve their reputation. Such initiatives do, however, shift the whole market up a gear, whether it likes it or not. Soon A-rated fridges will be the norm, and devices such as digital TV set-top boxes will be kept within strict energy-consumption limits.

The picture that has emerged of our TV-buying habits in the course of trying to impose energy labelling demonstrates perfectly why such labelling is needed, and why ultimately firm standards should be set. Left to the market, TVs are getting bigger and more energy-hungry all the time. The yearly running cost of a large plasma TV has been estimated at nearly £75, compared to £46 for a conventional one and £33 for an 'energy-accredited' TV. The bigger the set, and the better the picture quality, the more energy it consumes, with some households going down the 'home cinema' route with seventy-inch screens. Energy consumption is also being pushed up by the trend towards having two sets per household. It is like the fridge-in-the-garage phenomenon: if you can't be bothered to recycle it, plug it in somewhere else. In the run-up to the 2006 World Cup, sales of large flat-screen televisions showed an increase of 80% on the same week in 2005, and in 2010 it was sales of high-definition TVs that boomed, swelling the numbers of unloved tellies. I visited one recycling plant in 2009 which was still waiting to process the pile of unwanted TVs that had been fly-tipped after the 2008 pre-Olympic buying spree.

So efficiency standards are essential, as well as some limitation on size – surely there can't be any argument for allowing TVs to get bigger and bigger – and ways of making them easier to recycle. What would be much better again would be a way to upgrade the basic box as new functions and screen technologies come on stream. But all of this has to be in the context of some externally imposed standard of what a 'good' telly amounts to.

In the coming years, EU officials have plans to set energy-efficiency targets for TVs, computers, boilers, air conditioners, and a host of other products. They will then turn their attention to energy-*related* products, such as shower heads (because the more efficient the use of water in the shower head, the less hot water you will use). Even toilet cisterns may come in for scrutiny – if they contain an unnecessarily large quantity of cold water, they act as reverse radiators and suck the heat out of the bathroom.

So far, UK officials and politicians are fully behind these efforts. They estimate that changing products in these ways will save consumers £28 billion worth of energy by 2030, amounting to as much as 24 million tonnes of CO_2. Their efforts will quietly transform whole sectors of industry. The war on stuff's direct energy consumption will not on its own solve the energy crisis, but these standards have a greater importance. They set a precedent for political action which is central to achieving the vision expressed in this book – that governments can set standards for products in order to keep our economic activity within environmental limits. They also prove that we can design better products if we need to.

The next target for these standards would need to be embodied, or embedded, energy. According to David MacKay, the energy taken to *make and transport* stuff of all kinds (excluding food), if calculated on a personal basis, is nearly as much as the combined total for using cars and planes. As we saw in Chapter 10, we have started to get a handle on this through carbon footprinting, the process of working out how much energy (and thus invariably carbon, because most energy is presently from fossil fuel) is involved in every step of bringing us a product, as well as how much it takes up when we use it, and how much is involved in dealing with it as waste. As discussed earlier, it is also likely that the more we have 'outsourced' the production to other countries, the less efficient will be the use of energy, because of lack of development of their manufacturing systems. The next challenge is to use the intelligence contained in those footprints to make them smaller.

So both energy efficiency and embodied energy would be key goals of product standards, along with other qualities of products – water efficiency, recyclability, land use and others. But what about another solution to immediate energy needs – products that generate their own energy – as demonstrated by Evie and Ed?

Replacing the batteries

They are miraculous things, aren't they, these little packages of power? Just a few pence and light, music, captured images and all kinds of play are within our reach. For a bit more, we can buy electronic devices to connect us to the World Wide Web wherever we are, and run just about every aspect of our lives. None of this would be possible without the portable battery.

The battery is generally held to have been invented in 1800 by Alessandro Volta, an Italian physicist, although TV historian Adam Hart-Davis thinks that the Iraqis may have had a primitive version, the so-called 'Baghdad battery' some 2,000 years ago. Early batteries were made of copper and 'ions' in the form of material soaked in salt solution. The idea came from a chance discovery in 1791 by Luigi Galvani, an Italian scientist who was looking for the location of frogs' testicles, and discovered that dead frogs' legs moved when a spark was applied to them. The statically charged metal of the scalpel reacted with the ions in the frogs' legs, conducting electricity, which triggered the nervous response. Volta's innovation (thankfully) was to realise that salt-soaked material could be used instead of a dead frog, and if the metals and ionic materials were layered, chemical energy could be converted to electrical energy.

Batteries have come a long way since these rather cumbersome efforts – Volta's device was the size of a table lamp. A major breakthough was the use of lithium, a metal previously used to treat bi-polar disorder. Batteries based on lithium-ion and lithium-polymer are lighter and more compact than their predecessors, but still pack a lot of power, so they have enabled the explosion of portable electronic devices. Nonetheless, overall the humble battery has failed to keep pace with the improvement in the processing power of gadgets – the average annual gain in capacity of batteries is 6%, whereas the average annual gain in computer processing power has been estimated as 100%. If batteries were improving at this rate we could have batteries the size of a coin powering cars.

Batteries are also fiddly to collect and recycle – they have to be kept separate from household waste and go to specialist recycling facilities to separate the metals. So of the 600 million batteries used in the UK every year (an average of twenty-one for every household), only around 3% are presently recycled. Even the EU Batteries Directive, implemented in the UK in 2009, only requires the recycling of 25% of used batteries by 2012 and 45% by 2016. Hardly ambitious.

The truly green battery is one that doesn't exist in present form (i.e. removable and disposable). It would be integrated into products so not need changing, would have much greater longevity, and the product would be capable of easy disassembly and repair.

There are a few embryonic devices around already. I've had a solar-powered pocket calculator for years. You can buy tiny solar panels to power chargers for electrical gadgets, enabling you to get your emails on the beach, if you're that sad. They are also handy for keeping mobile phones alive while mountaineering, or providing light at an 'off-grid' campsite. They tend to be treated as a bit of a gimmick, but there is no reason that they shouldn't be improved with the right incentives.

The next most likely source of portable power is fuel cells. Fuel cells work like batteries in that they use chemical reactions to produce electricity, but with crucial differences. Batteries utilise the properties of some metals to give up bits of their atoms (the ions) – these are made to flow from one end of the battery to the other, making electricity as they go. In a single-use battery it is a one-way journey – when all the ions get to the other end, the battery is dead. In rechargeable batteries, an injection of mains electricity sends them all home again, to make the journey again and again. Eventually, though, after several hundred charges, they die of old age. In a fuel cell, rather than the largish quantities of metal, there is a small amount of platinum powder (though attempts are being made to produce catalysts from less expensive metals). When fed hydrogen from a tank, the powder splits the hydrogen atoms into positively charged ions and negatively charged electrons. At the same time, oxygen is introduced from the air, which also splits, and then reacts with the hydrogen ions and electrons to produce water. On the way, it helpfully produces electricity.

The fuel cell can do this indefinitely, with a fraction of the metal resources needed for a conventional battery. The only drawback is that it has to be fed hydrogen, which is obtained by reforming natural gas or by splitting water. The former is a fossil fuel, and both processes need energy. The advantage is that hydrogen is portable (if rather flammable) and can be made anywhere, so it could be made in places where the energy comes from a renewable source such as the sun, or hydroelectric (water) power.

Fuel cells have had decades of research and development and hundreds of millions of pounds of investment, but are only just reaching the market. They are already used to provide quiet, clean power for camper vans in Europe, and chargers for mobile phones and laptops are likely to the first portable applications – a methanol-powered mobile phone has been in development since 2008. Fuel cells have been slow to take over from conventional batteries because they are relatively complex and expensive to produce. The rising cost of fossil fuels may mean that the high cost of fuel-cell production and innovation may become commercially competitive, and the effort and expense that has previously gone into fossil-fuel technology and research could be orientated towards the growth and improvement (and lower cost) of fuel cells.

After fuel cells, the possibilities get more speculative, the kind of things that feature in my sons' 'cool futuristic stuff' genre of books, which offer some useful clues about product innovation. There are various forms of 'energy harvesting' – capturing the kinetic (movement) energy from people just walking around, or the energy from a running shoe hitting the ground. There is already a watch powered just by the movement of the watch on a wrist, and other likely devices are small phone chargers. Another clever 'harvesting' is piezoelectric, which is the energy given off when crystals are squeezed. This has already found an application in a hi-tech tennis racket – the crystals in the quartz fibres in the strings transmit electricity to a chip in the handle which stores it and fires it back to the strings again, causing the racket to stiffen more quickly and thus return the ball much faster.

None of these technologies is, as yet, looking like a dead cert to make metal batteries a thing of the past. As with all the other innovations described here, they won't necessarily be taken up simply because they're clever and better for the planet – they need clear market incentives. Since it's hard to 'pick winners' among these emerging technologies, the simplest course would be to announce the demise of disposable batteries, at a given date some years hence, and let the market innovate in response to that constraint. The justification for this would be a combination of the technical difficulties of recycling batteries, plus the logistics of collection. Portable power has been an amazing convenience, but in future it needs to combine convenience with renewable, low-impact technologies.

Fitting it all together

So what are the 'lash-ups', or ideal combinations, for energy? Large-scale renewables replacing current fossil fuels for electricity, feeding into an updated grid, which also takes and distributes the power generated by micro-generation from individual homes. The 'smart' grid will also work out where and when power is needed, and fire up or shut down household appliances appropriately. More advanced batteries will store surplus power until it is needed. At the same time, there will be a reduction of demand, particularly for fossil-fuel gas for heating, through better control of household comfort, as building standards require better designed and insulated buildings and smarter temperature controls. Wherever possible, household appliances and gadgets will generate their own power. There will be a more sophisticated plan for heat, as opposed to electricity, using waste heat from industrial processes, as well as biomass and biogas. All of this made even more effective by a horror of wasting energy (just as with food) which is instilled from infancy, backed up by very high energy charges for those who consume more than an 'average' amount, including industrial users. And if all this fails to balance demand with supply of renewable energy, there might need to be 'caps' on personal and industrial consumption, i.e. allowances for how much energy is available.

You will notice in all this that I have failed to mention coal or nuclear. Evie and Ed's world has coal, but not nuclear. This is because large areas of the world are sitting on extensive coal deposits, and although it would be good to think that they would develop renewable energy alternatives very quickly and leave those deposits alone, that is probably unrealistic. So clean coal technology is a must if only as a stopgap to more extensive deployment of renewables. It is another thing that should be possible but we just haven't thrown enough money at it yet. 'Carbon capture and storage' systems for coal-burning power stations are in development, which might prevent waste CO_2 emissions from contributing to global warming in the atmosphere. Renewed political commitment and funding are needed over the next few years if they are to make a difference in the next couple of decades.

As for nuclear, I really would like to do without it. Plenty of people I know in the environmental movement have changed their minds about nuclear power, and are now convinced that it can be done safely, and that it is a good low-carbon option, and that we will need it to complement other 'renewables'. There is plenty to read on this and I will leave you to draw your own conclusions, but I remain in the old-fashioned green camp of being unable to separate nuclear power from nuclear war. Nuclear power was developed in tandem with nuclear weapons because making the first can be a way to produce material for the second: plutonium. In my simple line of reasoning, the more countries that have access to nuclear power, the easier it is for countries that are so minded to develop bombs. International controls are getting better but they are not foolproof. Let's try harder with renewables.

17

Stemming the flow of stuff

Adhering to the principles we've discussed so far is a matter of both attitudinal and technological development – seeing things differently, but also having the available techniques and technologies to run them differently. But there is another dimension to Evie and Ed's world that is a necessary complement to these design developments. As we discussed in the first chapter, all reprocessing of materials costs energy, and even when that energy is renewable, it never comes completely 'free'. Wind turbines and solar panels have to be manufactured and installed, energy is lost in transmission, and storage devices such as batteries have their own material and energy costs. In the course of collection and reprocessing, a proportion of materials will inevitably be lost. All this means that the more materials and products can be reused in their present form, rather than reprocessed, the better. Rather than continual change, what that means (and it can be hard to get our heads around) is, in some circumstances, forgoing change.

'Reuse' sounds simple, but in a world of 'planned obsolescence' it is not the cultural norm. We are encouraged to change our stuff long before it is really worn out, either because simple repairs or upgrades are made impossible, or because we have been seduced by a new model. Academic Tim Cooper has calculated the discard rates for common household items, i.e. the % thrown out while still working – for computers 59%, for carpets and clothes

50%, for appliances and TVs 33%. Friends of the Earth put the spotlight on the turnover of mobile phones – although the amount of resources used for each phone has declined dramatically with technological improvements, the number of people using them has soared, and they are replaced on average once every eighteen months. That meant that as of 2005, an estimated 500 million 'obsolete' phones were sitting around, with less than 1% recycled.

Reuse takes a number of forms. The simplest is just not replacing something even if we have the fancy and the money to do so – sticking with that comfy sofa for another few years, or the rather out-of-date telly, or the classic coat. I kept one mobile phone so long that staff in phone shops laughed openly when I attempted to buy accessories for it, but resolve is needed here. The next simplest is passing stuff on to someone else who can use it, whether as gifts, through sites such as 'freecycle' (which matches things people want to get rid of with people who want them) or through second-hand commerce such as eBay, car boot sales, charity shops, the antiques trade or second-hand clothes markets.

Reuse is by far the best way to reduce the environmental impact of clothing and other textiles, given the resources involved in producing fibres and how relatively tricky they are to recycle. It is also part of a long tradition. There were the Florentine *regattieri* of the fifteenth century who traded the used, luxury textiles of the rich of Florence and made them accessible to humbler folk by remaking and reselling them. Textiles were at the apex of conspicuous consumption at the time – there were few other things to lavish money on, save, perhaps, furniture and silver. Sumptuous silks and brocades from exotic places were therefore investment pieces, with good resale value even when decades old. Similarly in Japan: for centuries kimono fabric, also usually silk, was recycled into new clothes, often in a patchwork of treasured fabrics. Sewn-together scraps were a mark of wealth and the ability to acquire a range of exotic textiles. A V&A exhibition in 2010 of elaborate English patchwork quilts, some of them still glowing with precious material and threads after three hundred years, demonstrated that fabrics really do last if looked after.

In this century, there is an encouraging and growing demand for second-hand textiles. Large quantities of clothes donated to charities are not sold in the UK (there is a home market for less than a quarter of what is donated) but are sold in bales to traders from all over the world. These are highly desirable goods – in the second-hand clothes markets of Zambia, for instance, Western used clothes are reinterpreted as 'new' and incorporated into Zambian dress in distinctive ways, often in combination with traditional textiles, rather then in slavish imitation of Western fashion.

So reuse of clothing has been, and still is, a strong feature of many cultures, and in the UK is making a comeback in the shape of eBay, together with a vogue for 'vintage' and jazzed-up charity shops made to imitate 'proper' shops. Even so, it has a long way to go to stem the tide of textiles going to landfill. What would help?

In Evie and Ed's world, the clothes kept in circulation are high quality. Today's clothes are not intended to last, so they are compromising the second-hand trade, whether destined for resale in the UK or exported. According to surveys, the proportion of reusable clothing in stuff coming to second-hand dealers has dropped from 65% to 40% over the last decade. To these dealers, a declining breed, quality of construction of a garment is more important than the label, whereas for the fashion trade the emphasis is the other way round. Here again pricing is crucial. Clothing has got steadily cheaper relative to incomes, but it is cheap for a reason. Manufacturing processes have moved to parts of the world where environmental and human conditions may be worse than in Europe, but labour is cheaper, and the desire for speed of delivery and constant change in the shops gives priority to quantity over quality. We buy more clothes than we need, discard them too readily, and given their poor quality we shouldn't be surprised if the only available destination is landfill. Higher prices seem to be the only solution – better to pay more for fewer clothes that last, and can be given to other people.

Antiques are another example of reuse – stuff kept in circulation not just for decades, but sometimes for centuries. Nigel Worboys of the Antiques are Green campaign uses the example of the Windsor chair, made by craftsmen in the woods and villages of England as long ago as the 1740s, still serviceable enough to be resold time and again, and still capable of being restored. Taking a slightly different tack, eBay has attempted to quantify how much carbon dioxide is saved by its 'peer to peer' operation, i.e. largely home-based buyers and sellers using the postal service, cutting out all the warehouse and transport infrastructure needed for conventional e-commerce.

Of course, in terms of overall environmental impact, what counts is whether these reused goods are genuinely substituting for new ones and reducing demand, or simply adding to the amount of stuff cluttering our homes. For 'reuse' to flourish to the extent that it supplants new material use (i.e. keeping things longer ourselves, or being able to pass or sell them to others) the ability to have things repaired is critical. In Evie and Ed's world, there is the ability to repair and remake almost anything. Many people will have acquired these skills themselves, since shorter working hours makes time available; for others this is their professional life. Both these options mean a new wave of skilled workers – those able to understand products and repair them, and those able to teach those skills to others. Design of products is still crucial, because they have to be designed for disassembly and repair.

Our present linear economy is going in entirely the opposite direction. Before cheap labour made the cost of products much lower overall, things like toasters and washing machines cost more relative to income, and repair was normal. For newer gadgets such as mobile phones, laptops and games consoles, it is well-nigh impossible to establish a repair culture – the skills simply do not exist in countries thousands of miles from where these things are assembled, and they are generally designed not to be messed with. The assumption is that we will want the newest one in six months anyway. There is also a worrying trend towards an aesthetic that assumes people don't want to be exposed to the 'workings' of things, that it is somehow vulgar or not 'modern'. Things that many people used to tackle themselves, like simple car repairs, are being denied to the amateur, in favour of paying a large amount for them to go to specialist dealers. American motorcycle repairman and political theorist Matthew Crawford argues that all this divorces us from stuff in a damaging way – we can no longer take the small satisfactions from having checked the oil or mended a clock. I would argue that it also makes 'material literacy' much harder, and makes us increasingly passive in the face of the people continually selling us things. We may be prepared to forgo the ability to understand and to mend when we're talking about something as multifunctional and useful as a smartphone, but when it comes to the toaster that can't be taken apart to put in a simple and inexpensive new element, there is a strong sense of powerlessness. In a similar vein, American sociologist Richard Sennett argues that to make or mend something for yourself is to understand it, identify with it, and probably cherish it much more than if purchased as part of a culture of gratification and disposability.

Reuse is, belatedly, becoming a more prominent part of public sector waste strategies – in 2010, the mayor of London, Boris Johnson, announced £8 million of funding for the London Reuse Network, a mix of local authorities and charities who would 'collect, store, refurbish and sell on everything from furniture, books, carpets and bikes through to cookers and fridges', aiming to divert 17,000 tonnes of reusable products from landfill over the first two years and over 1 million tonnes by 2015. That would be about a sixth of what is landfilled from London annually at the moment.

I am conscious writing this that it is very hard to give a plausible account of greater product longevity in a society that has been taught to continually move on to the next thing. But we already have 'old' things that we treasure – houses (what if everyone wanted a brand-new house?), classic cars, Aunt Mary's pearls. This is an area where Harvey Molotch's 'lash-ups' are critical – there has to be a felicitous melding of desirable design, acceptable price, available time (although taking things for repair needn't take longer than shopping for a new model), special skills and changed social meanings for all kinds of products. And for some products, the newer models may well be best – more energy-efficient, or deploying less hazardous substances, for instance. So we have to be helped to know what is the best thing – keeping some things while passing on others, or sending them for recycling where that is the best use of the materials. Standards for products that help to ensure some of these qualities of durability and repairability are part of achieving this, as I will argue further in Part 6. But values also have to play a role. We have to feel that newness is not the be-all and end-all for a lot of what is around us.

We could also share more. Design philosopher Victor Papanek suggested asking some fundamental questions every time we think of buying something, including 'Do I really need it?', 'Can I buy it second-hand?' and 'Can I rent it?' But the questions should also include 'Can I share it?', 'Can I borrow it?', 'Can we own it as a group?'

Papanek's survey of his 23-household street in Kansas revealed the ownership of fifty different models of vacuum cleaner, whereas he could make a case for just seven between that number of people.

18

Care with new promises

This is the principle that helps us to ensure that new technologies work to solve problems, rather than adding new ones. Two categories loom large in recent debates about what it means to exercise caution in the face of apparent technological advance – genetic modification and nanotechnology.

Genetic modification

Evie and Ed have the benefits of the products of a particular form of genetic modification. This is the kind of genetic modification that entails tinkering with the genetic code of microorganisms, either bacteria or microscopic fungi, and turning them into mini-factories for substances that we want to manufacture. This is a 'biorefinery' concept, and it is already used to make the enzymes that go into your washing powder, and is being researched as a way to make new types of plastic.

I learned quite a lot about genetic modification while serving on a series of government committees overseeing its regulation. I had been asked to sit on the first of these, a committee of mainly scientists, as a 'lay' member – someone chosen because they don't have a background in the science and therefore, in theory, won't have a leaning towards particular scientific ways of thinking about it. Put simply, I was there to ask the 'stupid' questions. The upside of this was the ability to challenge deeply held assumptions about the safety and efficacy of new technologies. The downside was, faced with an array of extraordinarily expert people, it took some time to muster the courage to do so.

After a while, however, I was quite comfortable asking stupid questions. What I learned about genetic modification was that the basic technology is very clever but quite simple – with the aid of enzymes that act like genetic 'scissors', pieces of the genetic code of one organism can be cut and pasted into the genome of another. Alternatively, genes can be transferred by putting them onto tiny particles of gold and firing them into host cells with a gun-like device. These techniques enable the transfer of traits that would not normally appear in the host organism, because they are crossing boundaries of species and even kingdom (for example, animal to plant). One of the first demonstrations of the power of the technology was putting the fluorescence genes from jellyfish into tobacco plants, causing them to glow in the dark. Little wonder that this started some public disquiet.

While the scissoring and firing are simple in concept, the tricky bit is identifying which genes do what, particularly when inserted into unfamiliar genomes. As a consequence, it's quite hard to come up with novel genetic combinations that actually do something useful, and so most genetic modifications of crops have been more mundane than the glowing tobacco. They have traits like resistance to herbicides (so that the whole field can be sprayed and the crop lives but the weeds die) and resistance to insects (the plants express toxins that act as a built-in pesticide). Although some of these have been popular with farmers outside Europe, there are relatively few commercial applications of GM crops, and their long-term impacts have been the subject of huge controversy.

Inside Europe, concerns about possible environmental

effects and the safety of GM foods have held up commercialisation. The environmental worries, substantiated by some seminal ecological trials held in the UK, have been around the effects of using herbicide resistance on a large scale and decreasing the amount of crop weeds available to support wildlife. There are also concerns around the possibility that genes from GM crops will transfer to wild relatives of the crop, leading to permanent changes in ecosystems. The essence of the concern is that we don't fully understand how genes work in the environment, and side effects may take a long time to be evident – but once they are, it may be too late to reverse them.

My time on these committees gave me a great respect for the scientists on all sides of the arguments, but I remain sceptical about our collective ability to understand and regulate something as fundamental as playing with the genetic code. I understand all the arguments about how in some senses GM is a development of age-old plant and animal breeding techniques, but I also agree with the critics who point out that conventional breeding never produced fluorescent tobacco. I don't buy the argument that GM is needed to feed the world – there are many ways to increase global food supplies (not least by reducing waste, as discussed earlier) before GM crops need come to the fore. The overwhelming flavour of the public debate was not only that people didn't like the sound of GM, but that they couldn't see why it was *necessary*.

However, I am prepared to concede that there are some specific instances where it might be a useful tool to have in the armoury. One that is worth debating is using GM to create disease-resistant bananas. The fact that bananas are sterile 'clones' of each other and therefore can't be cross-bred with better bananas to resist the ravages of diseases, including the black sigatoka fungus that is devastating the agriculture of many countries that rely on exporting bananas for their livelihood, means that GM might be justified. The inability of bananas to reproduce by normal means should also ensure that any newly inserted genes can't spread.

The other possible place I see for GM is in the biorefinery – in contained facilities, producing substances that are not themselves genetically modified but are the products of genetically modified microbes. This is done by inserting the genes that produce the substance of interest (for instance, the fat-digesting enzymes present in fungal organisms) into faster-growing microorganisms that act as a kind of biological factory. They reproduce and make lots of the enzyme in the course of multiplying, which can then be harvested and the GM bugs destroyed. All this takes place in sealed vats, with all wastes thoroughly treated and no GM material reaching the outside world. I went to see a Danish facility where our 'biological' washing-powder enzymes are produced in this way, helping to dissolve those annoying greasy stains. The staff took round a whole party of environmentalists, and after the tour and detailed explanation of the techniques, none of us could find much to complain about. That made a change.

So I feel comfortable positing Evie and Ed's various materials as the products of genetic modification combined with clever chemistry, enabling the substitution of materials such as oil-based plastics and valuable paper pulp with new materials made from plant waste products. But all of it takes place under lock and key. I remain to be convinced that we know enough about how GM crops work to be comfortable with widespread releases into the environment.

Nanotechnology

Nanotechnology means the use of very, very small particles in industrial applications – we're talking about bits of material measuring a millionth of a millimetre.

Materials get interesting at the nanoscale – they can completely change their character, so that benign aluminium, for instance, becomes an explosive substance when reduced to nanoparticles. They are invaluable in manufacturing because they can be applied as ultra-thin films of material. This makes them key to the production of silicon chips, where the repeated layering of minute quantities of metal and the formation of microscopic gold wiring is what has enabled the growth in processing power of our electronic gadgets. This 'thin film technology' also enables layers of metal to be incorporated into packaging, like the aluminium that forms an antiseptic layer in juice cartons. Nanoparticles are gradually being incorporated into all kinds of goods without fanfare – one survey suggests that as of 2010, nanomaterials are present in wound dressings, water filters, towels, cycling shorts, razors, badminton rackets, cosmetics, sunscreen, self-cleaning glass, socks, car polish and tennis balls.

Nanotech has been held up by some commentators as a route to a variety of environmental solutions, including ultra-thin films of gel on windows to reduce heat loss, higher-performing photovoltaic cells, and better ways of storing hydrogen, seen as key to cleaner transport. One of the most appealing ideas I've come across is mimicking the colours of butterfly wings. The iridescent colour is not pigment, it is tiny particles reflecting light at different frequencies, and the same effect might be achieved by man-made nanoparticles. This could ultimately replace dyes and be a more environmentally benign means of generating colour.

There is also the quest to produce very strong materials, and here carbon 'nanotubes' show great promise, including their use as 'space tethers'. The first time I saw this written down I thought it was a joke. But no, there is an entire scientific community devoted to space tethers. The process of making the carbon strings has been described as making carbon into smoke, so that the particles entangle and hold hands, creating 'elastic smoke' that can be woven into fibre. One application for this idea might be in constructing 'space elevators' to take satellites and space-station materials up into orbit without the need for rockets. There might also be ways of transmitting solar energy collected in space back down to earth, or using the motion of the elevator itself to generate power.

However, for these 'smart materials' to be truly smart, they need to do more than enable fantastic new technologies. They also need to be smart in ways that counter, not exacerbate, the linear economy. They need to be low-energy-consuming, low-water, non-toxic, non-persistent and recyclable. It is not at all clear that the industry is developing according to these criteria. Present means of manufacturing nanomaterials are highly energy-intensive because of the amount of purification that has to be done to isolate the particles – carbon nanoparticles can be between two and a hundred times more energy-intensive to make than aluminium, for instance. Applications like microchip production are also very water-intensive, because the demand for purity of the tiny layers of metal means repeated washing, often with solvents as well, which end up as pollutants in the waste water and have to be removed in further energy-intensive processes. Concerns have been expressed from many commentators about the potential toxicity of nanoparticles to people – not least because they are small enough to cross cell membranes. There is also uncertainty about their ability to persist in the environment as pollutants, particularly since materials at nanoscale behave differently to materials at usual scale. That is precisely what makes these materials interesting, but it also means that existing information on toxicity and persistence might be of little help in assessing long-term risk.

So the potential for new applications of nanomaterials to solve environmental problems has to be weighed against some already apparent downsides, and there is a clear case for guiding further development with standards that help to ensure more efficient manufacturing processes, and rules for how these materials are allowed to enter the environment and what happens when their useful lives are over.

Some scientists talk about 'smart' materials as an enticing part of the future – smart meaning that they respond to conditions. Some of these employ nanotechnologies, others are a more conventional scale. But just as with the composite and nanomaterial agendas, they need to subscribe to the cradle-to-cradle principles in their design. Otherwise, the supposedly 'smart' might actually be rather stupid.

Part 6

From Here to There

19

Brighton and Beyond

March 2010. A digger-type machine loads clawfuls of stuff from the floor of a building the size of an aircraft hangar onto a conveyor belt. The stuff has come from those among the residents of Brighton who have been diligent enough to use their recycling boxes rather than putting everything into black bags. The contents of the boxes come to this state-of-the art facility; the black bags go next door to a smaller shed before being sent to their final destination, an energy recovery plant.

The conveyor belt is the first stage of taking useful things from household garbage in what is called a 'materials recovery facility' (MRF). The plant is fed by collections of 'commingled recyclables', which means that paper, cardboard, plastic and metal can be put into one recycling box and collected fortnightly by the local council. Householders are given a list of what can or can't go into the box, although, as I later witness, they don't always rigidly stick to it.

At the other end of the conveyor belt, a small cabin suspended in the middle of the hall is home to two men who pick off the most obvious 'reject' material. Most of this is loose plastic bags, or recyclables tied up in plastic bags, which people persist in doing despite being told when given the boxes that plastic bags are neither necessary nor welcome. After that, the conveyor passes through a huge perforated drum like the insides of a giant's washing machine, which revolves slowly, churning rubbish. Inside the drum, the mix of stuff is separated into what Jeannette, my guide, calls 'flats and rounds'. The larger pieces of flat paper and card spill out of the end of the drum, the bottles and cans and smaller bits and pieces of paper and plastic fall through the holes. Seventy-five per cent by weight of what comes through the plant is paper, which is removed at several different stages of the process, so there are conveyors carrying paper in all directions. It's like a reverse image of those old films showing newsprint snaking through a printing works – but here the paper is going back from being newspapers and magazines towards, eventually, being blank paper again.

The end of the process for the paper is in another picking cabin where a line of men removes the last of the inappropriate materials. They include plastic bottles that have escaped into this stream, yet more plastic bags, and plenty of crisp packets (no type of crisp packet is presently recyclable). The mountains of paper are then rammed into bales and wired up, before being stacked neatly at the side of the building, ready to go to the paper mill. Jeannette tells me that it's quite hard to get it wrong with paper – just about everything that a household could throw out is welcome at the paper mill, where it can be turned into newspapers. Big catalogues and telephone directories can be a problem, as their glue-filled spines present a challenge to the machinery, but with care even this can be overcome.

The conveyor carrying the bottles, cans and bits takes them to a screen made of spinning discs, which lets any paper and bottle tops fall out and the bottles and cans continue as a cleaner stream. Next, a large magnet snatches the steel cans and drops them into a bin below. A gizmo called an 'eddy current separator', like a magnet in reverse, creates a force field that causes the aluminium cans to jump out and head for their own dedicated bin. The really clever, and fairly recent, bit of kit is the 'optical sorter'. This is the technology that magically recognises different types of plastic, which it does by knowing their 'signature' wavelength when bombarded with infrared light, and little puffs of air then direct each type into different bins. At the moment it does two types – PET and HDPE – but in future could do more, provided someone wants to buy more types of recovered plastic. Like the paper, the cans and plastic bottles end up in large multi-coloured cubes, like giant 3-D modern art.

I take the most interest in the conveyors carrying the rejected matter – this is the stuff that can't be sold and ends up next door in the shed of shame. This MRF has a lower rejection rate than many, at only 7% of the stuff that comes in, measured in tonnage. But that's still 2,000 tonnes out of the 30,000 tonnes that comes through the door every year. Something from it is 'recovered' because it is sent to a plant where it is burned to produce electricity – but it would be better if all those materials were somehow re-usable.

The rejections are partly, but not entirely, down to the generally willing citizens of Brighton not paying quite enough attention to their list of things that can or can't go in the box. Nowhere does it say glass – glass gets smashed up and mucks up the recovered materials badly – yet still bits of glass turn up. To be fair, some councils do take glass in their commingled collections, so some people may be going on the advice of friends and family in other localities. In fact, many of the 'out-of-spec' items are entirely forgivable from the perspective of the harried householder. Coat hangers are plastic, aren't they? Yes, but not the right kind. Milk, juice and soup cartons are cardboard? Yes, but they also contain layers of plastic and sometimes thin metal foils, so they need specialist processing which presently only takes place in Sweden. Jeannette has a box in her office of stuff that fails the plant (she, and I, see it that way round, rather than the plant failing the materials). Much of what is in her box is complicated plastic and metal combinations, like cosmetics dispensers, and fussy plastic and paper mixes that crop up in many types of 'luxury' packaging. There are also the crisp packets, which tend to be a sandwich of metal between layers of plastic, and indeed any packaging material of this type such as cat-food pouches. There are plastics that are unidentifiable and therefore automatically suspect. But the ones that really annoy Jeannette are the ones with those 'recyclable' logos on them, the triangle of arrows with a number inside. Margarine and ice-cream tubs, for instance, carry a number 5, which means that they are polypropylene (PP). Although looking and feeling very similar to PET (plastic no.1) or HDPE (plastic no.2), and being technically recyclable, PP is not yet reclaimed in sufficient quantities to find a market. The same is true of polystyrene, often labelled as plastic no. 6. Of course, if people see that logo, they are tempted to stick them in the box, whatever the list says. Jeannette thinks it would be better to label only those commonly recycled, as many retailers have started to do for packaging materials.

The element of the reject pile that really dismays me is the itsy-bitsy stuff, since I am guilty of including this kind of thing in my own recycling box. So shredded paper, train tickets, receipts, and all kinds of tiny bits of paper cannot be captured by the machinery and end up with the second-class, non-recyclable materials. I make a mental note to put these in the compost container instead, where they will happily contribute their carbon to the mix, although I'm not quite sure what will happen to the metal strips on the train tickets.

The useful stuff is more heartening. Bales of paper off to paper mills to be turned into newsprint, and card to be turned into boxes and similar products. PET and HDPE plastic go off to companies that turn it into food grade plastic packaging again, or reprocess it into polymer products ranging from fleece jackets to street furniture. Cans go to steel and aluminium smelters, some to be turned back into cans. There are some emerging 'closed loops' in here, stuff going round in a circle and being turned back into what it was before. All the stuff that comes in is either recovered as materials or burned to produce energy, which means that there is a market for what, until ten years ago, was considered waste and cost money to dispose of.

The site cost in excess of £16 million to build. Why is it here? Did the company see its future solely in selling post-consumer materials to a few places in the UK and beyond? No, it's a major waste-treatment company, contracted by the local council, upon whom the EU and in turn the UK government have placed legal targets for diverting stuff from landfill, as well as telling it to meet targets for recycling of household waste. Those requirements have triggered a government 'private finance initiative' to meet the investment cost. This is materials recovery as advanced waste disposal – done as well as present circumstances allow, but it's still materials recovery in its infancy.

It brings me hard up against the many questions rehearsed in this book. Do people care enough to change their material habits, if some can't be bothered to read the list of recyclables properly, or in some places, even take advantage of the recycling facilities at all? How can we redesign products to go through this kind of process more seamlessly, more efficiently, so as to get more useful stuff out of it? How do we ensure a steady market for recovered material – indeed, make recovered materials the norm rather than the exception, as in Evie and Ed's world? And why concentrate on households, when so much more and very similar stuff is chucked out by businesses every year in the UK?

The fridge mountain

Not long after my visit to the MRF, I get the opportunity to visit some facilities where the materials recovery is driven more by markets than by waste disposal policies alone, but still not completely. Metals recycling plants deal with our chucked-out cars, fridges, washing machines, as well as computers, phones, games consoles and a host of other gadgets, collectively known as 'waste electronics and electrical equipment', or WEEE. To get a sense of how much of this type of stuff we use in an average lifetime, take a look at the WEEEman sculpture at the Eden Project.

Markets for recycled metals have always been vigorous, for two main reasons – metals are expensive to produce relative to materials like paper and plastic, and they are relatively easy to reclaim. They can be melted down and mixed with new sources of the metal, and small quantities of unwanted accompanying material are vaporised in the process. That means that even tiny amounts of precious metals embedded in printed circuit boards, for instance, can be smelted out, with the bits of plastic and other stuff disappearing up the chimney.

There has always been good money in the stuff we tend to associate with scrap-metal dealers – steel from cars, copper from old plumbing, aluminium from machinery and even drinks cans. But this trade has been boosted in the last decade by European laws requiring that 95% of cars (by weight) must be reused or recycled by 2015, and more recently that each European country must reach a recycling rate per person per year of around 4kg for consumer electronics (about the weight of my laptop, steam iron and toaster combined). That target has been quite easily met, and so from 2016, the proposal is annual collection of at least 65% by weight of all the WEEE put on the market in the previous two years, a likely personal average of around 20kg. Again, some countries will do this easily, but others, including the UK, will have to raise their game. The key thing about both these directives, and the forthcoming one that will require recycling of batteries, is that manufacturers have to help meet the costs of recovery.

The process for reclaiming metal is roughly the same whether it's a transit van, a TV or a toaster. Cars and other big scrap are fed through a 'shredder'. More accurately, this is a giant mincer, since the scrap is pounded with unimaginable brutality by huge hammers on a revolving drum and then somehow forced through a giant sieve. It then goes through several more pieces of machinery to separate different metals, emerging as uniformly sized nuggets of valuable stuff about the size of an orange.

Fridges require special treatment because of the ozone-depleting CFC chemicals that are in the compressing units (the motorised bits that create the cold air) and in the foam that insulates the body of the fridge. Fridges have to be carefully drained of their chemicals and oils, and then the rest is put into a sealed unit, five or six fridges at a time, and pounded to bits with flying chains to extract the steel and usable plastic. As described in Chapter 16, there is a surge in TV purchase (and concomitant old-TV disposal) in the weeks before a major broadcast sports-fest such as the World Cup or Olympics. Apparently the same is true of fridges, presumably in order to ensure state-of-the-art beer storage. If only the beer drinkers could see the row upon row of discarded fridges, awaiting their flaying.

The stuff we take to local authority sites and put in the WEEE bin is similarly minced up. The resulting flakes resemble large, garishly coloured confetti because of all the plastic casings, but you can see that it is interlaced with metal. This mix is less than half metal by volume, so in the next step magnets extract the steel, and what is left goes through an eddy current separator similar to the one in Jeannette's plant that nabs the aluminium cans. This one is set to claim different metals as they jump different heights in response to the current. The resulting 'eddy dump' pile of mixed metals is still very colourful, given other stuff still attached. After the next process, which takes out a lot of the plastic (which can also be sent for recycling) the pile is rather less colourful. It still doesn't look like anything in particular, but is full of small quantities of the precious metals in circuit boards (gold, silver and copper) and could be worth several hundred pounds per tonne, depending on the state of the markets. This is sold to processors worldwide who will smelt it to recover the valuable metals. What is left at the very end is about 10% by volume of completely unusable 'fluff' – dirty brown dusty stuff that is mainly packaging, old Hoover bags and other ancillary material.

The WEEE operation is similar to that at Jeannette's plant: the extraction of usable materials from mixed-up post-consumer waste. Like the MRF in Brighton, it is driven by legally binding recycling targets. Local authorities collect the WEEE on behalf of producers and then sell it on for processing. Without the targets, this kind of collection would not be economic.

WEEE regulations bear on companies rather than on local authorities, because they are framed under a type of European law known as 'producer responsibility' instruments. In theory, this should create a much more direct incentive to design for recycling, because making recycling difficult downstream should work back to the company as increased cost. Unfortunately, this incentive has been diluted by the creation of 'compliance schemes'. These are separate companies who take on the burden of showing that recycling targets are being met by signing contracts with recyclers on behalf of a number of producing companies. While this simplifies the process for the individual companies, it also puts them back at arm's length from the consequences of their designs.

So for the metal recyclers, just like in Brighton, life would be much simpler and profits higher if the metals were easier to separate from the products, and there were fewer materials overall to deal with. Cars contain over forty types of plastic, many of them stuck to each other, which makes it harder to reclaim economic quantities of plastic. There is concern that future car design might go in the wrong direction, opting increasingly for composite materials such as high-strength carbon fibre and metal combinations. They have a good safety profile, but given their mixed-materials nature, they are unlikely to be easy to recycle.

Both the End of Life Vehicles Directive and the WEEE Directive are supposed to influence design, but the recycling companies have not seen much evidence of this so far. At the same time, WEEE is growing faster than the rest of the domestic stream, as new gadgets are dangled in front of us all the time. The new stuff coming on stream takes no account of how effectively it can be recycled so it is the recycling companies that have to invest in new kit to keep up with the product manufacturers, rather than the other way round. You might expect car-makers and designers of electronic and electrical equipment to have made some kind of approach to metal recyclers to explore how they might influence and even support the technologies that will deal with their products at 'end of life'. But no – nothing at present either requires or encourages such a tie-up.

There is also a 'producer responsibility' regime for packaging, which should prompt design improvements that take things out of Jeannette's box of non-recyclables. Unfortunately, here also the dilution effect of compliance schemes and the lack of ambitious targets have meant only limited progress. Most packaging manufacturers have concentrated on taking weight out of packaging, which helps them to meet targets based on tonnages, and also saves money. This is good in the sense that it saves materials and gives a better carbon fooprint (i.e. less energy taken to manufacture and transport materials), but less good in the sense that it has given rise to a lot of very lightweight but unrecyclable packaging. Really, we need it to be both. Meanwhile, because the recovery targets can be met by taking the 'easily' recyclable materials out by means of plants such as Jeannette's, the complicated mixes of materials and bitty unrecyclable stuff like bottle tops and plastic strips, have gone largely unaddressed. As with the orbiting space junk, the WEEE at the recyclers and the abandoned fridges, nobody thought about 'end of life' when dreaming their design dreams, and nobody has been mandated to sort it out at source.

The current mechanics and market difficulties of recycling provide ample illustration of just how far we have taken stuff. We take nature's raw materials, transform them in ways that nature could never achieve, and put them in places they never normally reach. It has huge benefits, but it has disbenefits too, both environmental and human ones. The market economies that have evolved globally, and are now intricately interconnected, take little account of damage to the environment in the way they price goods and services, and take insufficient interest in keeping stuff in productive use. So environmental damage is the inevitable consequence of 'wealth creation' as currently understood, and that damage threatens to undermine our ability to create wealth in future. Like many other aspects of the global economy, that 'undermining' will not be equal – it will hit some people and places harder than others. Exactly how it will affect any of us, including those of us fortunate to live affluent lives, is unclear.

What matters is intent, and whether, if the right people change their intent, we can reverse the most damaging trends. Judith Rees, a very eminent academic and writer on natural resources, quotes Machiavelli: 'There is nothing more difficult to carry out, nor more doubtful of success, nor more dangerous to handle, than to initiate a new order of things.' Yet new orders have been achieved by the determined – reformation of churches, the widespread institution of democracy, votes for women, abolition of slavery. More recently, banning smoking in public counts as a significant act of leadership by the governments of Scotland, England and Wales in 2006 and 2007. On the environmental agenda, the example to cling to is the unprecedented swiftness of the global agreement to phase out ozone-destroying chemicals. It can be done. So what happened to make Evie and Ed inhabit, and indeed shape, a different world?

20

Humans grow up

Evie and Ed were born at the very end of the twentieth century, and were eight and ten respectively when the credit crunch of 2008 hit. Both sets of parents were reasonably, but not very, well off – both working, paying for a house, affording consumables such as cars and holidays. In the economic downturn of 2008, all of them feared for their jobs, but only Ed's dad lost his. He made the decision to stay at home and look after the children because his wife enjoyed her job more, and could earn just as much. It saved on childcare costs and gave him a good deal of satisfaction overall, although he was quite often bored and lonely, especially when the children went to secondary school. Ed's parents moved out of London to a village where life was more sociable, and offered just as much scope for informal business such as selling secondhand goods on Internet auction sites and doing DIY jobs for people. Apart from a few friends who had had to change jobs, and the news that economic growth had slowed right down, life didn't seem all that much different.

Evie's family stayed in their home city in the north of England. Before the recession, the area had begun to show signs of regeneration after decades of lagging behind the south, and boasted the same coffee-shop chains and shiny shopping malls as the rest of the country. The recession put a brake on that, and the gap between the haves and have-nots widened again. Nevertheless, for Evie's parents, that was where their family lived, so they stayed put. With less money to spend, Evie's and Ed's parents, along with many of their friends and relations, started to look around them and wonder if they really needed as much as they had been accustomed to buying. Christmases strewn with battery-powered toys, gadgets and oceans of shiny wrapping paper were all good fun, but it was so much more stuff than when they were children. There was also so much cheap, out-of-season food – again, enjoyable, but not really necessary. And having twice as much stuff around as when they were growing up hadn't made their children demonstrably twice as happy – in fact, probably the contrary. Now, they went shopping less often (spending the time cooking or playing games instead) and started to make things last longer. They took advantage of the few remaining repair shops, or learned new skills such as sewing and electronics. Besides providing satisfaction, these were ways to involve the children in producing some of the basics of everyday life. They felt virtuous, but, importantly, they didn't feel unusual or alone.

At roughly the same time, the UK government decided to play an active role in European efforts to set standards for products, which would start to make transparent the amount of energy, water and materials involved in their production, and also to steer all products towards being eventually fully recyclable. The beginnings of this idea came through the government's own 'green procurement' drive, where suppliers to public sector organisations had to be much clearer about the implications of their products, and to accept that they might not be allowed to supply the government if their products didn't have the right credentials. Procurement also started to favour products that could verify they had come from 'sustainable sources' – first timber, and then metals and minerals – ensuring that right back at the stage where resources were extracted there was a measure of control over the environmental and social conditions in play. Taking the step from procurement standards to legal standards on all products was made politically possible by signals from some important retail businesses that their customers supported these principles. The companies also started to believe that taking the environment out of the realms of competitive behaviour by making *all* businesses act would make them more popular. This might have resulted from sheer good sense, but may also have been a consequence of some much vaunted 'green' claims seriously misfiring to the extent that shareholders became irritated. Either way, from a mixture of the 'green procurement' pressure and leading businesses' own efforts, the consistency and transparency of companies' treatment of resources improved vastly over a five-year period, so that when legal standards for products came along, they weren't that onerous to put into practice. The government also helped the process by redirecting science budgets and funding a lot of research and discussion with businesses about how best to measure and judge what was good and bad about any given product, and about what needed to be done.

More surprising was the UK government's victory in Europe on being allowed to set rates of VAT different to those levied in other member states. This was then used to give preference to more 'environmentally sustainable' products, by taxing resource-intensive products much more steeply. This differential taxation was made possible by the information gathered in the process of setting product standards. Together, these measures sent strong signals to businesses that the hidden environmental costs of consumption could no longer be ignored.

These measures had very swift results. Some products disappeared, others emerged. Once the pattern of product development became clear (and it is astonishing how fast businesses can adapt and innovate once the direction of travel is certain), it was much easier to plan the infrastructure needed for recovery, reuse and repro-cessing of materials. In some cases, such as electrical and electronic goods, cars, washing machines, furniture and clothes, the businesses producing them (or the people importing them from other countries) were given direct responsibility for getting them back and making the best use of the materials. In others, recovery was organised by retailers, whose 'supermarket' sites became 'resource parks', accepting products and materials back from customers, sometimes paying for them, and thus catering for the cycling of materials through the economy as well as for their dispersal to customers. By the time Evie and Ed were teenagers, they both lived in areas offering this kind of system, and to earn extra pocket money, one of their jobs in their respective homes was loading up the (electric) car with the household's recyclables to take back to the supermarket.

Local authorities retained a role for more difficult and dangerous wastes, but were generously funded to provide that service through the product taxes. So despite its 'Big Brother' connotations, government had not designed the 'universal recycling system' at all, either local or central. It was the product of cooperation between large retailers, who had seen the writing on the wall in terms of the escalating costs of losing materials from the economic system, whether from their own higher disposal charges, lost custom because of higher VAT on bad products, or customer anger if materials could not be recycled. In other cases, sale of the product was replaced by sale of the service – such as clothing or tool libraries, or leasing furniture rather than owning it.

For energy, the policy of paying incentives for renewable energy, whether generated by energy companies at large sites or at home by solar panels and other gadgets, gradually transformed the market. People began to demand devices that made energy rather than consuming it, to plug into their 'smart grids' and earn money. They got even more of a premium if they kept their own usage low, or diverted surplus energy to the new, highly efficient batteries installed underground in built-up areas. These could be used to charge electric cars and bikes, or for backup in the event of grid failure. These measures to 'redistribute' energy helped to overcome the Jevons Paradox of people reacting to having cleaner and/or cheaper energy by simply using more of it. The household-scale biogas plants utilising household food waste and sewage took a long time to develop and fine-tune, but eventually freed many areas from dependence on fossil-fuel gases.

None of this happened overnight, and none of it was without cost. Some things got more expensive, others were cheaper (because of the incentives to save resources during manufacturing, and the greater availability and security of reclaimed materials). For some things that were felt to have high environmental impact but still be essential for certain uses, such as some metals, 'caps' or limits were set on the amount of new extraction, making use of recycled materials essential. The government spent money on public education, on persuading businesses these moves were in their best interests, and on rigorous enforcement of rules.

Internationally, there was considerable argument over how far all these moves should be allowed to affect trade and economic growth. The picture was mixed – the 'older' economies stayed on a path of very low growth, but since people were reporting greater levels of well-being, this didn't matter too much politically. The brave moves to condition the environmental impact of products had been accompanied by some equally bold moves to cut consumption of products damaging to health, including alcohol, tobacco, sugar and saturated fats, by imposing even greater taxes. Benefits had been cut but working hours had been made shorter and more flexible, enabling more people to be in work even if on lower salaries. By the time Evie and Ed were both looking for work, twenty-five-hour weeks were the norm. The profits of financial institutions were heavily taxed and the money was used to support welfare and volunteering packages for those really unable to work. The newer, high-growth economies were still expanding vigorously, but were reinvesting a great deal in developing better products in order to keep up with the new standards. Once better living conditions had been secured for the majority of their citizens, they were able to accept lower rates of growth.

Perhaps the hardest battle was to reform the terms of trade with some of the poorest countries on the planet. Many of the wealthier countries (which by now included China, Russia and India – in fact, most of Asia and quite a lot of South America) argued that African nations' problems were of their own making for failing to tackle corruption and establish good government. Some more enlightened countries (including the UK) argued that even if that were the case, the way that terms of trade had been drawn up by the powerful post-war economies did poorer countries no favours, and by making it very hard for them to be equal players in the globalised economy, probably contributed to their failure to govern themselves well. Eventually, African products were given better access to world markets, which meant that more money could be earned and kept in those countries, reducing the need for aid. Aid budgets began to be kept for emergency relief after natural disasters (and climate change brought about increased storms and flooding), but they were also channelled into resettling people away from vulnerable areas as well as investing in better systems of capturing and storing water. Most important, a vastly increased international fund was established for ensuring that all women who wanted contraception could easily access it. That, together with improved incomes, started to bring down family sizes and stabilised world population more quickly than had been hoped. Evie and Ed met when she was volunteering for a research programme into sustainable fish farming for Africa, and he was training family-planning professionals ...

I'm sure many people will find this a ridiculously optimistic account of the future – and I wouldn't blame them – but I remain unabashed. I don't know any other way to appeal to a sense of the possible. So let's imagine it is possible. What can we infer from this picture about who should be taking the lead?

Will science save us?

We are clever, but our cleverness is not entirely a modern phenomenon, and neither does it surface with consistent speed or results. Our view of technological development tends to be a very selective one, as historian David Edgerton has made clear. We see a linear progression from one great innovation to the next – steam engines, motor cars, aircraft, computers – when in reality the road is more bumpy, and pockmarked with failure. Innovation means simply novelty – we notice the new thing that succeeds, that is taken up, but fail to notice all the unsuccessful inventions and ideas that fall by the wayside. Some things are taken up long after they are first invented, some old inventions coexist with more modern ones. We also tend to ignore the more mundane but nevertheless life-changing products and technologies in favour of the exciting ones that seem to define an era. Chlorine bleach, for instance, which is a cheap, effective antiseptic, may have environmental downsides but has saved many lives by preventing the spread of infections. What is important is not so much how and where things arose, but how we use them, mixing and matching the old and new, re-inventing where needed, shedding all the time, as in Harvey's Molotch's 'lash-ups'.

So, for instance, we sometimes think of ourselves as in a post-manufacturing age, yet Chinese success can be put down to the efficient deployment of established manufacturing technology, which has far outstripped the growth of the much more 'innovative' Japanese market. The rise of the humble goods container, a simple if enormous metal box, and the increasing use of the age-old technology of shipping, has probably made a greater contribution to escalating resource use than many other more 'modern' technologies.

The lesson of these insights for dealing with environmental problems is that we mustn't always wait for a magic technological bullet to turn up. It's an easy route for politicians to claim that problems are so intractable that they can only be solved by something not yet invented, which in turn provides the rationale for delays in using government policy to address the issues. Nothing illustrates this better than responses to global warming, where despite continual political statements about how important it is to solve, everyone seems to be waiting for the technologies that will deliver the cheap, convenient, totally environmentally benign renewable energy to turn up on a plate. Few politicians seem able to grasp that not only is there a suite of perfectly good renewable technologies to deploy already if the incentives were correctly aligned, but that if these technologies are to evolve and improve, access to established, dirty technologies must be restricted. As Edgerton and others point out, innovation is largely a reaction to scarcity, otherwise people carry on doing what they already know about. Oil companies spending big money to extract very dirty oil from the tar sands in the Canadian wilderness rather than going all-out to produce the same energy from solar power is a good example of this.

The Jevons paradox is another limitation on the role of technology and innovation. Raising the prices of resources does lead to innovation to promote efficiency, but if that in turn simply lowers costs in a way that promotes greater consumption of the kind we have already, it does more harm than good. The basis of consumption has to change towards new, low-impact technologies such as renewable energy, water conservation technologies and ways of designing for recycling. To achieve that requires more than simply making resources more expensive; it requires specific rules or other incentives to do things differently – in other words, that *direct* the right kinds of innovation and technology. As Jared Diamond observes, to get science and technology to contribute towards a better environment, we have first to stop it contributing to any further damage.

Don't ask, won't get

'Green innovation forcing' was a strong theme of one of the many government advisory committees that I have been involved with. The Commission on Environmental Markets and Economic Performance (CEMEP) was created by the New Labour government in 2006, at the instigation of Gordon Brown when Chancellor. Roughly summarised, if not officially stated in this way, its brief was: 'If we have to do all this green stuff, how can we reap the biggest benefits for UK businesses?' The commission comprised fourteen members drawn from companies, academia and trade unions, with me as the 'NGO' representative. On top of that we had David Miliband (then Secretary of State for the Environment) and Alistair Darling (then head of the Department of Trade and Industry) and a couple of more junior ministers. Miliband and Darling were supposed to be joint chairs of the commission, although their diaries didn't often permit them to perform that role in the flesh. Despite being somewhat hampered by the lack of a solid brief, including no clear definition of 'UK businesses' (owned in the UK, or just operating in the UK?) or 'UK competitiveness' (how do you pin down 'benefits' to the UK when there is no such thing as UK plc?), we had some very productive discussions. We ended up being clear that however much innovation is going on, not much of it goes on in response to environmental pressures. That is because innovation tends to come from scarcity imposed either by price (as I have emphasised throughout this book, prices largely don't reflect environmental damage) or by legislation. Where there is environmental taxation (such as the land-fill tax) or environmental legislation, there are examples of innovation. The gradual shift of investment from land-fill to recycling facilities is one example. The state of California's decrees demanding that there must be cleaner car exhausts, kick-starting the hydrogen car market, is another. But these examples are relatively few. Without government measures creating a steadily increasing price for carbon, for instance, *as well as* measures to set the standards for products, improvement of the environ-mental credentials of goods and services cannot be expected to be consistent or coherent.

The government's response to CEMEP's ideas has been patchy. Some ideas have been taken up – most notably the pressure to introduce feed-in tariffs for renewable energy. But the ideas about looking beyond the carbon agenda to materials and to water, and creating opportu-nities for new directions in British business by specifying better products, have yet to take a hold in the corridors of power. As the next section will argue, until they do, we won't get very far.

Leaders need to lead

Governments do already set frameworks for the market. They may try to sound as if they do it with a very light touch, but that is rarely true. What counts is where their priorities lie. For instance, the UK is proud of having the best workplace safety record in world, but we would be nowhere on the environment without the intervention of the EU. So I go to recycling plants in continental Europe where the materials recovery rates are excellent, but there is a slightly greater danger of being run over by a forklift truck. Their people come here and admire the discipline around the hard hats and vehicle control, but tut at the piles of stuff heading for landfill. Both are products of state intervention, but reflect different preoccupations.

We are fortunate to have joined the European Community. It is the only example globally of a multinational lawmaking body that largely works, and quite often makes the laws stick. So for the kind of things that we're looking for to shape Evie and Ed's world, the EU is a good place to start. At the same time, the UK has been proactive not just on the climate change agenda, but also on the idea of product policy, which gives some grounds for hoping that a UK government could take the lead on making these ideas work. At the end of the day, though, *someone* has to lead, and be prepared to negotiate the undoubtedly numerous political obstacles that will crop up on the way. As we've seen, consumers can shift markets, but only in piecemeal ways. Everyone can have ideas about how things should change – pressure groups, consumers, designers, progressive businesses – but altering the status quo in any widespread way requires aligning all these interests to a degree that is unlikely without some external authority.

'Green' products, or every product the 'green' choice?

Products are already subject to rules and regulations, many of which we're not conscious of. There are chemicals that are banned from foods, rules that govern the hygiene of the factories where food is produced, and regulations that govern use of sell-by dates. Products must not be dangerous to the user, so they must conform to standards for testing and certification. Seeds used by gardeners must conform to international standards on how many of a packet can be expected to germinate. Most of these diktats have to do with quality and safety; they are to project us consumers from the product failing to fulfil its promise or turning round and biting us. Surely protecting the environment from products is just as important in the longer term, and just as valid a part of production and consumption?

Most rules are enacted on the basis that companies on their own are not going to move to put such constraints in place, and so the market is failing to deliver some important aspect of public interest. In many cases, companies have no quarrel with the overall goals or do anything other than adapt to them once the rules are in place. For some companies, regulations give competitive advantage, because they eliminate competition from those less scrupulous in their operations. In a few cases, companies themselves proposing constraints on their sector have prompted regulations. This can be in order to develop in new directions or simply to protect reputation. One of my first jobs at Green Alliance was to organise a coalition of very unlikely bedfellows (including Friends of the Earth, the Women's Institute, and the British Agrochemicals Association) to lobby for revision of pesticide laws, because the continued presence on the market of older pesticides that hadn't had full safety appraisals was seen by leaders in the industry as likely to undermine public confidence in their products.

The product standards picture

What kind of standards would we want to set for products? We would want them to be designed to use minimal energy, and wherever possible generate energy, either to power themselves or to feed energy into a grid. We would want them to be very conscious of the water needed to produce them and any water needed for their use, taking account of where that water comes from and the other demands that might be on it. We would certainly want standards to ensure that products are designed for easy recovery and recycling. We might also want standards to encourage minimal use of, or replacement of, certain materials, either because of the implications to the natural environment of getting them, or because long-term supplies are in doubt. We would certainly want standards about how long a product should last – long enough to get a decent life out of the materials and enable upgrading and repeated reuse, but not so long that an energy- or water-using product can't be replaced by more efficient models as they come along.

How would we do this? There are a variety of options. As discussed in Chapter 16, energy-efficiency standards are already being put in place by the European Union. They work by specifying that, in future, your television, set-top box, computer or boiler can only use a set amount of energy per hour of operation. Finding that magic figure is a matter of intense haggling between European bureaucrats and industry representatives, but they get there in the end. It also involves a group of national and Brussels-based NGOs, including Green Alliance, collectively known as the Coolproducts campaign, who try to counteract the inevitable watering down of ambition throughout the lengthy negotiation process. The products, the environment and our energy bills will be all the better for it – before their intervention, the market for some products was going in entirely the wrong direction, producing ever more energy-hungry appliances.

These standards will start to govern how products are designed in order to reduce the energy and water involved in their *use*. The next step is to look at how much energy and water is involved in their *manufacture*, and set standards for that as well. This is beginning with the development of carbon footprints, which give companies a methodology for understanding where and how energy is used the whole way along the supply chain for a product. Remember the carbon-labelled crisps mentioned earlier? It's not that these crisps are especially low carbon, it is just that they were one of the first products to be subjected to this kind of analysis, and then labelled with a final carbon figure. The value of this to consumers is dubious unless every single brand of crisps is labelled and people really do want to choose their crisps on this basis. But the value to the manufacturer is huge. It enabled the company to identify the carbon 'hot spots' in their operation and do something about them. With more of this kind of work, it should be possible to map the hot spots and look for solutions. Then it should be possible to set standards that reflect what the whole of that industry could achieve.

What of materials? The fact that some products are recyclable and others not is a major limiting factor to reversing the linear economy. Recyclability should be quite a simple standard to set in theory, although in practice it would entail massive change for some products and materials. The only way to do this would be through setting the goal with a long time horizon, probably a minimum of ten years, to allow manufacturers to innovate around the goal. For some products, the constituent materials may be already perfectly recyclable, but the way they are joined together is the problem – this is the case, for instance, with packaging that uses glued-together layers of different materials, or products made from different plastics, or products that join metals with other materials such as plastics and paper. In other cases, it is the materials themselves that aren't easily recyclable – the lightweight plastics that are hard to recover in sufficient quantities to find a market, or the composite materials discussed earlier. In both cases, standards aimed at improving recyclability could start by specifying greater recycled content of some products – plastic bottles, paper packaging, some uses of metals – because this would begin to pull more reclaimed materials through the system, make them the norm rather than the exception, and encourage manufacturers to design for recycling. It will never be possible to have 100% of products recycled – there will always be some loss of materials in the system – but we could aim for 60% on all fronts and radically reduce the amount of virgin materials taken, as well as the amount of materials wasted. For the really lightweight or bitty stuff, compostability would be the better route, as discussed in Part 5. This would be the standard for those products.

There is one initiative already in existence that might help to point the way to the future. Not of the product as such, but of the packaging. Since 2008, by means of a voluntary code adopted by retailers, the majority of packages carry a logo to indicate the recyclability of the packaging materials. This system was thrashed out by retailers in conjunction with local authorities in response to criticisms that the diverse collection schemes offered by different localities were confusing to consumers, and that there were inconsistencies in the materials accepted. So the logos indicate whether a material is 'widely recycled' (i.e. accepted by more than 65% of local authorities), or whether consumers should 'check with local authority' (i.e. between 15 and 65% of authorities accept it), or that the material is 'not presently recycled' (i.e. less than 15% of authorities collect it). The last category can mean either that it is generally uneconomic to collect, or that it is not physically recyclable. So, for instance, a ready meal in a PET tray with a thin film of plastic over the top and a paper sleeve might declare the tray and the sleeve as 'widely recycled' but the film as not recyclable, although not the reasons for this. The value of this system has been not only to guide householders as to which bin to put stuff in, but to influence the design of packaging materials, since retailers (and thus suppliers to the retailers) were reluctant to use the 'not presently recycled' labels more than absolutely necessary. If this labelling system and general trend were to spread to products as well as packaging, it could have wide impact.

As of 2010, an interesting approach is taking shape in France. The French government has passed legislation requiring that, from 1 January 2011, all products will have to carry a set of 'environmental indicators'. These are likely to include an estimate of the quantity of greenhouse gases released during the main stages of the product's life cycle (growing, manufacturing, packaging, and transporting and retailing), although this comes with the disclaimer that it is valid only in metropolitan France. There are indications that this would include showing where on a scale of carbon impact the product sits – high, medium or low. This would go some way to solving the problem described above for the crisps – not knowing how 'good' or 'bad' the carbon figure is. There will also be a label indicating how much of the packaging can be recycled *if* the consumer separates it properly. The methodology for these labels is being developed by government-backed agencies in conjunction with businesses, and will become the national standard for arriving at these 'indicators'. The fact that *all* products will be labelled means that the indicators might offer genuine choices, and the fact that the methodology is a standard, widely accepted approach should help everyone in coming to conclusions about which products are better or worse than others. What then happens to the 'worst' ones is an open question. Will consumers actively select against them? Will companies in the supply chain weed them out of their own accord because they make them look bad? Or will the government have all the information it needs to start to set standards for products?

We also need to set standards that reflect best practice

in the way materials are extracted, in ways that enable companies to make the right choices further up the supply chain. This is the only way to condition the impacts of primary activities such as mining, quarrying and logging, most of which go on outside our shores. We have the beginnings of such standards in certification schemes such as FSC for timber, or the early days of metal stewardship as demonstrated by the Responsible Jewellery Council, both of which aim to encourage best practice and thus deny the less scrupulous a market. If it were legally required that certain products came from certified sources, this system would be put on a much firmer footing. At the moment, anyone selling paper has a choice as to whether it comes from FSC or other certified sources, or not. In future, the standard for paper would stipulate that it had to come from such sources – again, with a long enough lead time to ensure that there are enough certified sources available.

The government as 'intelligent customer'

There is a way to move towards product standards that would help to move markets in the shorter term, while also building up valuable experience as to how standards might work in practice. This is through the dull-sounding mechanism of 'public procurement', or the government buying things for its own use. It is often trotted out that the NHS has the spending power of a small African state, and that UK public bodies are responsible for handing over around £6 billion per year to companies for the goods and services they need to keep the country ticking along. The leverage in that £6 billion ought to be considerable, in the sense that if the government says it wants 100% recycled paper or FSC-certified timber, it is such an important customer that it ought to be able to have it, yesterday.

The reality is less straightforward. That £6billion is distributed among many different budgets and individual buyers, all with a host of criteria, not always consistently applied, as to how to spend public money. All of this has to be achieved while still ensuring 'value for money', which generally translates as lowest possible cost. 'Sustainability' has recently been tacked on to this list, but as with many of the other criteria, it is a lot to expect people who are trained as buyers to fully take on board what this should mean in relation to every single product.

As a result, the Department for Environment, Food and Rural Affairs has produced guidance to direct buyers across the public sector towards environmental 'quick wins'. They include buying recycled paper, FSC-certified timber and abandoning bottled water.

It is not too hard to envisage the expansion of this idea in two directions. One is to start to say that, eventually, the government will only buy products with the same environmental credentials as those presently considered the 'best'. In that way, the 'quick wins' gradually become the norm and, de facto, set a standard that anyone who wants to supply to government has to meet. The other direction is for government to specify the environmental choices that it would like to have in future, even if those choices don't presently exist. By promising to buy a 'greener' product if one can be invented, the public sector provides the impetus for companies to invest in changing what they do.

The Prison Service did this for prison mattresses. Dismayed at how often mattresses had to be thrown out, they asked potential suppliers to propose solutions that would be more durable and recyclable, and, if possible, 'zero waste'. I was asked to help to assess the tenders, and most of the proposals were a vast improvement on how the process was being managed at the time. This way of going about things is called 'forward commitment procurement' – you invent it, we'll buy it.

The other advantage of using procurement as a weapon in the war against waste is that for many suppliers, public bodies may be such large clients that if they have to develop new solutions for them, they may as well produce it for their other customers as well, and by then they are on the way to transforming the whole market for that product.

Pricing out waste

There is another important lever that any government can use in order to encourage better products and get waste out of the economic system. This is the power to set taxes. In the UK, most public money is raised through taxing our labour (income tax) or our properties (council tax). A proportion is raised through taxing products (VAT) which doesn't discriminate particular products, other than to exempt things deemed 'essentials' i.e. food, children's clothes, medicines, etc. Very little is raised by taxing damage to the environment, despite several decades of commentary from academics and NGOs on how much more sensible this would be – 'tax bads, not goods'.

The few environmental taxes we have include the landfill tax, which has been very important in signalling to the waste industry that landfill is not the future. It started at a low rate of £6 per tonne (the basic rate for using landfill was at the time around £30 per tonne, so the extra £6 was not going to make much difference). However, successive Chancellors accepted the arguments from Green Alliance and other NGOs, as well as the waste industry itself, that alternatives to landfill such as recycling and composting plants would only be economic when landfill ceased to compete. The tax has therefore been increased over a number of years, with future levels announced with plenty of notice, so that it will reach £72 per tonne in 2013. Now, waste companies are able to make a case for building expensive plants like the MRF in Brighton.

The landfill tax is an important, but fairly blunt, instrument. It directs stuff away from landfill, but makes no judgement about where it should go instead, or how products should be designed to maxmise the chance of reuse or recycling. Green Alliance has long proposed the introduction of product levies. This would be a way of favouring 'greener' products in the marketplace, for instance by charging VAT in a way that differentiates between good and bad products. Unfortunately, tinkering with VAT is something that we have to do in agreement with our EU neighbours, since to have wildly different rates of product taxes in different countries could distort trade. This may seem an impossible task given that the EU has now expanded to twenty-seven countries, but in 2008 President Sarkozy of France together with Gordon Brown did at least put the issue of differential VAT on the EU agenda.

There is yet another point at which products and materials could be taxed to alter their environmental profiles. It is upstream again from waste treatment or products, at the points at which materials are extracted, manufactured and then imported. Taxing at these stages could mean a comprehensive carbon tax, levied on energy as it is used, and preferably by all countries around the world so that all global trade would feel the effects of it. A carbon tax would do a lot to change products and their wastefulness, because it would make less economic the materials that take most energy to produce, the least efficient manufacturing processes and the least desirable forms of transport – those based on greatest use of fossil fuels. It could even affect the wasteful use of water, because water takes energy to pump and to clean.

If particular materials are a concern, these could also be subject to international levies similar to a carbon tax. We have a precedent for a 'raw materials tax' in the UK's Aggregates Levy, which imposes a tax on the gravels and sands taken from sensitive areas such as certain coastal areas and the seabed. This does two things: encourages greater use of recycled aggregates, because they become relatively cheaper; and also contributes to a pot of money to restore the environment. According to government figures, the levy has reduced the sale of 'virgin' aggregates by around 18 million tonnes between 2001 and 2005, and led to an estimated increase in the use of recycled aggregate by nearly 6 million tonnes, a quarter of the total aggregate used. Levies on raw materials could be imposed as materials cross borders, with the money recycled, by agreement, to conservation projects in their country of origin.

Making these kinds of taxes and levies work on an international basis is no small task. It would require unprecedented agreement between countries, and the establishment of well-resourced and fully trusted institutions to administer them. The world has not done brilliantly on this front so far, with the 2009 climate talks in Copenhagen ending in a disappointment for those who expected a binding deal. There are some positive precedents, however. The global agreement known as the Montreal Protocol, which all but phased out ozone-depleting chemicals, is a rare example of political agreement and action inside anything other than a glacial timescale. If we want to move away from the linear economic model on a global scale, we have to believe that global political cooperation is possible – indeed, we have to expect it from those who lead us.

Businesses

You may feel from the list of expectations above – product standards, better buying practices, new forms of taxation and global agreements – that I am expecting rather a lot of governments. This is true. Only they can lead comprehensive change, but they can't do it without cooperation from all the other actors in the economy. What is the role of businesses in securing the future?

Businesses provide things, and they innovate, but mostly they try to satisfy the needs and whims of us as consumers. They are, of course, highly influential in the creation and shaping of those 'needs' and whims, but it takes two to tango – we can't absolve ourselves completely of the demand for 'stuff'. They also cater for government as a customer, which is, as I've already argued, where some of the big opportunities lie for rewarding companies with progressive products and policies, and gradually setting a tone for all business behaviour.

In my view, the chief responsibilities of businesses are these. They should help governments to understand the realities of business life and the competitive pressures under which they operate, including the attitudes of the consumers they serve, and point out why it is often hard for businesses to act on environmental issues unilaterally. They should be prepared to negotiate and inform but then accept the rules proposed by governments, as long as they provide 'a level playing field' and can be phased in with enough time to innovate. They should refrain from being 'ethically competitive', unless they can rigorously demonstrate the substance behind their claims to be better. They should provide information to the consumer, preferably in accordance with standardised protocols, so that there is no confusion and no plethora of different labels and unfounded claims.

Green Alliance has been working with businesses who want to debate these ideas. They want to improve the environmental profile of the products and services they offer their customers, but they are hampered by the need to keep costs down and to remain 'competitive', i.e. not losing customers to the less switched-on companies who carry on with 'business as usual'. As we've seen throughout this book, just because doing the right thing in environmental terms appeals, it doesn't make it pay. These companies know they need governments to help set the frameworks within which they all operate, and so they have been taking part in a continuing discussion about what exactly the government's role should be. They may not all agree that *legal* as opposed to voluntary standards are needed, or agree on where fiscal incentives might be applied, but they do know that someone has to lead an effort to raise the collective game, and they are prepared to play their part. And if businesses can work together to provide a radically greener generation of products *without* the facilitation of government, that would be a welcome and groundbreaking result.

Return to sender – the ultimate producer responsibility

Measures such as product standards and pricing of environmentally damaging products would usher in an age of much greater business action on the impacts of the stuff around us. But does that really equate with taking more responsibility? 'Producer responsibility' is a term often used loosely to mean any passing of the waste buck from public authorities, such as our local councils, back towards the people who make the waste in the first place. In the examples of the EU directives on cars, waste electronics and batteries, it simply means making producers more responsible for the costs than they used to be, because they have to pay compliance schemes to ensure that recycling targets are met. But at its logical extreme, producer responsibility would mean requiring companies to *organise for themselves* the passage of their products through the economy from 'cradle to cradle'. So they would have to think about how materials would get back into use before putting anything on the market, and would have to ensure it was all done with minimum energy and water. The products and materials would never stop being theirs, and they would have total responsibility and liability for them.

This would be the ultimate lever on design – businesses would have the freedom to design their own solutions (and I mean design in the broadest sense, not just the bit that makes the product aesthetically pleasing), rather than work to common standards negotiated with public bodies. That would also get over the problem of the negotiation process lagging behind the pace of change in business and not taking sufficient account of innovation. But there would be no escape from the responsibility for the whole life of the product.

Such a system would require governments to legislate to create this way of organising things, but, having done so, they could leave companies to it. The enforcement would be through everyone's scrutiny of everything on market – if the products on offer didn't accord with customers' sense of what constitutes 'responsible', the companies would be taken to task.

Which is the better route? Carefully negotiated and probably incremental standards, or a radical shift in responsibility (backed up by citizen enforcement) followed by a rapid burst of innovation and creativity? The latter sounds more appealing in principle, but would be politically a lot more difficult to achieve, since it would involve increased costs for many businesses. It would also require a much more environmentally literate body of consumers, whose responsibility it would be to spot those products not expressing the new values. There is a danger that it might anyway slip back to being a 'standards setting' exercise in order to be sure of meeting defined goals.

The only way it could possibly work, to secure the necessary degree of change, would be through the kind of 'radical transparency' and disclosure advocated by Daniel Goleman and the 'mass balance' work pioneered by Biffaward (mentioned in Chapter 3), so that all the bad things about current products and the good things done by companies (and most important, their competitors) were entirely open and up for debate. Even then, it's a big ask that accountability for progress will be sufficient. On the other hand, the solutions coming forward from business might be that much more interesting and progressive.

At some point soon the business community needs to choose which route it wants to go down – extended producer responsibility or the negotiation of detailed standards – and then go down it with good grace.

Us

I hope I have demonstrated that leaving the environmental credentials of products to the market, and the well-meaning choices of 'green' consumers, has not so far substantially changed products. It is not that green consumers aren't important – they have been crucial in making companies look at themselves, and listen to messages about what people want to see on the shelves. The trouble is, there is insufficient pressure to re-evaluate *everything* on the shelves. Supermarkets in particular are schizophrenic places – you can buy the very good and the very bad, often side by side. It mirrors our own schizophrenia – wanting to do good, but deciding to buy according to other criteria when it suits us, be it price, appearance or peer-group pressure. Neither side has taken the lead. With few exceptions, retailers have not constrained our choices, and we have not voted with our feet in sufficient numbers for the few choices that do exist. To expect that would be expecting too much – green products are presently niche, often expensive and don't always perform. We are in a catch-22.

There is another reason why the power of the green consumer is necessarily limited. We can exercise choice in relation to individual products, but there is little we can do to influence whole systems. These are the product of the evolution of our various ways of making and disposing of things, with these two sets of activity taking place largely in their own bubbles. It is impossible for individual consumers to decide how we manage the nutrient cycle, for instance, because it involves big infra-structure decisions about how we treat sewage and waste water, and our reliance on mined fertilisers. A large-scale shift to compostable materials is also beyond the buying choices of householders.

However, we should all make stronger demands. I argued in Chapter 10 that consumers have limited power to change mainstream products because the 'ethical' choice is catered for – the trick is to make the 'ethical' mainstream. We can seek to do this by demanding the right things, all the time and very loudly. Not everyone has time or energy for this, but for those who do, it is vital to up the ante so that politi-cians feel pressure to step in and firm it up. Most of all, we should not give mixed signals – moan about destruc-tion of the environment and then buy cheap, dubious products. Pressure groups can be an effective proxy for consumer and/or citizen concern – but only if they main-tain pressure and don't continually skip about from agenda to agenda. At the end of the day, though, without some common, consistently applied standards for products, none of this will completely do the job we need.

We must also be prepared to accept incentives (or penalties – they are two sides of the same coin) for certain behaviours. One of these would be for throwing out usable things rather than taking easily available routes to putting them back into circulation. Couched as an 'incentive', householders would be rewarded according to the quan-tity of recyclable goods placed in the containers they are given. Couched as a 'penalty', householders would be charged for waste *not* finding its way into those containers – presumably a function of either ignorance (in which case an incentive or penalty might quickly foster knowledge) or sheer laziness. I am as guilty as the next person of not consistently putting the right things in the right bins, but I'm sure I would think about it a lot harder if there was a financial dimension.

We should also be prepared to accept 'constraints' – no longer having the choice to buy some products. Retailers do a certain amount of 'choice editing' on our behalf already. In 2008 B&Q made a big play of not selling patio heaters, following a decision made by Wyevale garden centres the previous year. At present, these examples are few, but in future a combination of product standards and increased consumer demand for better stuff could completely change the complexion of what is on the shelves. Whether that results in *fewer* products overall is an open question – I suspect that corporate ingenuity would kick in and give us variations on a theme, it's just that the theme would have a more acceptable profile in the first place.

Capping consumption?

We've considered a range of things that we as individuals might need to embrace. They include having products 'edited' out or radically changed by a combination of government and business action. They also include accepting incentives to cooperate with important systems such as recycling materials properly and shifting to renewable forms of energy, as well as a shift in values so that we try to stave off the lure of continual novelty. Will all this be enough to keep our activities within the limits discussed in Chapter 8? We can't presently answer this question, because we don't know what those limits actually are. We know, pretty much as a matter of logic, that we cannot go on as we are, ad infinitum. But we don't know where the limits lie because we don't know with any level of detail or certainty what the real crunch points are (although we do know that more than two degrees of global warming is likely to be problematic). We have the choice, as outlined in Chapter 8, to let the environment draw the boundaries for us, by gradually or suddenly being unable to supply what we need – fresh water, a stable climate, clean air, viable soils, food, timber, and every- thing else. Or we can attempt to draw those boundaries ourselves, and limit resource use and pollution in defined areas, and hope that our calculations are near enough to keep us safe. If we do the latter, it could be the first step towards defining a 'fair' share of the planet's resource for each of us, wherever we live.

Tim Jackson has set out one of the most recent and compelling series of steps towards 'ecological macro- economics' in his book *Prosperity without Growth*. It sounds daunting, but in essence it means a system of economics that takes account of, and works within, strict limits on the use of resources, and on the polluting consequences of using those resources. For Jackson, a key part of this is to shed our current preoccupation with 'growth', which is always assumed necessary to keep the economy stable and achieve full employment. This is partly because the amount of people needed to run a developed country goes down as technology improves – you only have to think of how few people work in a car plant (alongside the many robots) compared to the legions of workers a few decades ago. If the economy isn't growing, there is no new work to make up for the work now done by the robots (or outsourced to developing countries). But what the future economy needs is actually *more* labour-inten- sive, low-material activities so that more people can be kept in work, but the resources they use up are fewer (Jackson gives as examples of low-impact activities, yoga lessons, gardening and hairdressing, knowing that he is open to accusations of promoting a 'yurt-based' economy). Another option is to reduce working hours to share the available work around more equitably. He freely admits that it's not clear that an economy can function with this kind of work as the norm, but it is essential to explore how it might.

At the same time, it is essential for us all to embrace the idea that 'prosperity' is not necessarily synonymous with material wealth. As Robert Kennedy famously said, 'Gross National Product ... measures everything except that which makes life worthwhile.' There are other ways to feel prosperous, what philosopher Kate Soper has termed 'alternative hedonism'. This might describe the satisfactions in Evie and Ed's world – shorter working hours freeing up time to spend with children, with friends, cooking, gardening, making things, mending things. It is possible to believe in prospering socially, intellectually and having more fun, flourishing in every aspect of our lives – apart from that which dictates that we must pursue the relentless acquisition of goods.

Reconfiguring the economy along the lines that Jackson suggests is the biggest challenge the human race has yet faced. As he states, it will require much greater 'ecological literacy' on the part of our leaders, as well as for all of us. Schools are increasingly trying to play their part in this, but overall there is astonishingly little drive to make us all understand just what is at stake when we stick stubbornly to the idea of evermore growth. We are back to the gap I identified in the first chapter – the secret life of stuff and the secret extent of its consequences.

The Eden effect – environmental 'literacy'

In an ideal world, we would acquire 'materials literacy' and a concern for the world while we are still children, and carry it through to our adult decision-making in a seamless fashion. This is the motivating force behind the

Eden Project in Cornwall. As the sign at the entrance says, it provides education but doesn't feel like school, it puts champagne in the veins, and it points to a positive future. It is a garden, a place of entertainment and a mission, and it is simply beautiful. I am biased since I have been a non-executive director of the project since 2005, but it is still the place I most look forward to arriving at, with the sole exception of home.

I first went to Eden in 2002, two years after its opening. Nothing compares with walking out of the entrance hall for the first time to look into the pit that is Eden's home – once a china clay quarry, now one of the most spectacular gardens in the world.

Eden is best known for having 'the world's largest indoor rainforest', an astonishing collection of tropical plants that were already large when planted, but which after ten years are straining at the seams of a dome that resembles (and was indeed inspired by) a captured soap bubble. All are planted in soil that had to be manufactured from the china clay waste products together with compost made nearby. The specimens are there not just to be admired, but to tell a story – the story of our dependence on plants. Wandering around, I am always enchanted by the art and craft woven into the tapestry of the many shades of green. Dominating the entrance to the Humid Tropics Biome is a mock-up of the enormous hull of a boat, painted green to symbolise green trade. Giant bamboo is made into a bamboo house of the kind traditionally built in South-East Asia, demonstrating its strength and vital role in the economy. Cocoa plants are interlaced with stories about chocolate. A large drawer unit of spices brings home the diversity of products from the forest and the history of their exploitation, and the spices are there to feel and smell.

In the outdoor gardens, more stories are told – plants that give food, plants that give dye, plants that make fabrics, as well as plants bred simply to delight. It goes beyond plants – at the very centre of the site there is a sculpture made of WEEE – one person's lifetime consumption of fridges, washing machines, TVs and many, many bits of things with dangling wires fashioned into a goggle-eyed monster. I have watched people stop and read the narrative attached to it, and then shake their heads in disbelief. Diners at the cafe are asked to separate their waste into recycling bins as part of Eden's 'waste neutral' programme, and informed that their food waste will be turned into compost for the garden, and the other recyclables sent away for reprocessing. The 'neutral' bit of waste neutral describes Eden's policy of buying in recycled goods as well as sending recyclables out.

To me, all this is magic, but then I understand the intent and I have been exposed to a large part of the story already through my work. What effect does Eden have on those who just come for a good day out, or are fleeing a Cornish beach when it turns from golden to chilly? As well as reaching passing visitors, Eden also runs schools programmes in which children are more actively engaged in the stories, so it's important to know what they take from it.

Eden has invested in high-quality analysis of how far its 'messaging' has an impact. As a consequence, the Eden staff know that the blend of direct experience of the natural world with carefully thought through messages, all put together with strong art and design, are very successful at getting adults and children alike to think more about the environment. Children draw a predictable picture of a 'rainforest' before they go into the biome – tree, monkey, banana. They draw much more diverse and detailed ones afterwards. They understand a bit more but, most important, they enjoy.

Not everyone can go to the Eden Project, however much I wish they could. Eden is developing ways of taking this blend of nature, science, art and theatre to a wider audience, so that more people have a way into the basics of the material world. It is an approach that other educators could learn from. Champions of 'environmental education', however worthy the enterprise, don't tend to fire much passion in their audiences. Eden does. If the magic could be somehow injected into places as diverse as nursery schools, professional institutes, the civil service and businesses themselves, we would be a lot further ahead. Already the Eden Project's 'green talent' programme is helping young people to understand the nature of the environmental challenge, and consider how they can use their talents and skills to contribute to a more positive future. I would like to think that Evie and Ed came here as children, and that it helped them to shape that future.

So we have reached the end of the secret life of stuff. Of course, it is only a glimpse. There is much, much more to be uncovered. But I hope it makes you feel a bit more powerful. Not just in the sense of wanting to flex your 'green consumer' muscles and select the currently most 'green' or 'ethical' products, which remains an important way to send a signal, and should not be abandoned. I also want you to be empowered by realising that we need our choices conditioned for us, that we should be given no option but to buy good stuff, and that such an approach is a lot more advanced than having unchallenged liberty to play fast and loose with the living world. This need not mean a diminution of well-being – in fact, quite the contrary. I am certain that we can enjoy different things, in different ways, and still live within our means on a planetary scale. I am looking forward to it.

Notes

Chapter 1

p. 5 'disposable drinks glasses' – *Dragons' Den* (2008), Series 6, Episode 3, BBC2.

p. 6 'Waste electronics' – NetRegs (2010). Meet targets for recovering and recycling WEEE (updated 30 June 2010). Available at: http://www.netregs.gov.uk/netregs/topics/WEEE/110599.aspx. According to NetRegs, around 50% (av. weight) of the materials in this category are actually recycled.

p. 7 'grams of plastic' – Light Emotions (2009) 'Glow Flute' Technical Datasheet. Available at: http://www.light emotions.com/home/images/documents/datasheet_glow_flutes_en.pdf

p. 7 'glass consumed in the UK' – UK Glass Manufacture (2008) A Mass Balance Study. Available at: http://www.brit glass.org.uk/Files/Mass_Balance_Final.pdf [accessed on 27 July 2010.] p.3.

Chapter 3

p. 14 'William Rathje' – Rathje, W., and Cullen, M. (2001), *Rubbish! The Archaeology of Garbage*, USA: University of Arizona Press.

p. 15 'clean their dens' – Girling, R. (2005), *Rubbish – Dirt on our Hands and Crisis Ahead*, London: Eden Project Books, p.14.

p. 15 (GDP) and waste' – Cyclope/Veolia Environmental Services (2009) *From Waste To Resource: an Abstract of World Waste Survey 2009*.

p. 17 'litter is left there' – Gurubacharya, B. (2006), 'Mountaineers prepare for clean-up mission on Everest', *Guardian*, 6 March.

p. 17 'junk in space' – Liou, J-C., and Johnson, N. L. (2006), 'Risks in Space from Orbiting Debris', *Science*, 20 January, Vol. 311, No. 5759, pp. 340–1.

p. 17 'space-person poo' – Mullane, M. (2006), *Riding Rockets: The Outrageous Tales of a Space Shuttle Astronaut*, Scribner, p. 76. The idea emerges from Arthur C. Clarke, who wrote 'Toilets of the Gods. Or: The Colonisation of Space', *Ad Astra*. Reproduced at: www.astrobiology.com/adastra/clarke.html It is a sci-fi conceit – but one that is based on poor waste management on early space missions, which has subsequently been improved.

p. 20 'plenty of sources' – Resource Futures (2009) *Municipal Waste Composition: A Review of Municipal Waste Component Analyses*, Defra project WR0119, March 2009.

p. 20 'produces dangerous chemicals' – see for instance the advice at http://www.milton-keynes.gov.uk/environmental-health/DisplayArticle.asp?ID=29284

p. 21 'generating electricity' – Hogg. Dominic (2006), *A Changing Climate for Energy from Waste?* Final Report for Friends of the Earth.

p. 22 'increased consumption' – Cooper, T. (2009), 'War on Waste? The Politics of Waste and Recycling in Post-War Britain, 1950–1975', *Capitalism Nature Socialism*, Vol. 20, lss. 4, pp. 53–72.

p. 22 'indulge itself' – Stobart, J. (2008), *Spend, Spend, Spend: A History of Shopping*, Gloucestershire: The History Press, p. 213.

p. 23 '30 million tonnes' – Cooper, T. (2007), History & Policy Memorandum, submitted to the House of Commons Environment, Food and Rural Affairs Select Committee inquiry: 'A New Waste Strategy for England'. Available at:

http://www.historyandpolicy.org/docs/waste_select_committee.pdf

p. 23 'new state of affairs' – Flintoff, F. (1969), 'Refuse: The Volume Explosion', *Public Cleansing*, July, p. 309. quoted in Cooper (2009).

p. 23 'disposable nappies' – Parfitt, J. (2010), Lifting the Lid – 'The First 140 years of the UK dustbin' in *Resource* Sep–Oct 2010, Number 55, p. 25.

p. 23 'back into the system' – Boulding, K. (1978), *Ecodynamics: A new theory of societal evolution*, USA: Sage Publications, Inc., p. 296. 'On the whole, the biosphere deals with the problem of pollution by recycling. The excrement of one animal, for instance, is the food of another. The carbon dioxide that animals excrete, plants ingest and turn back into oxygen, which the animals breathe in and use and then again turn into carbon dioxide. Nitrogen passes through plants into the atmosphere where it is fixed in the soil by nitrogen-fixing bacteria associated with other plants and is again available for protein formation. Carbon is constantly recycled from the atmosphere to the soil, through the biosphere, into the atmosphere, and the soil again. Many other elements are recycled in a similar way. The human race indeed is almost the only living being that has invented a linear economy, distributed into dumps or flushed down to the oceans or burned in the atmosphere. This is obviously a temporary arrangement, but exactly how temporary is a little hard to say. Ultimately, if the human race is to survive, it must develop a cyclical economy in which all materials are obtained from the great reservoirs – the air, the soil, the sea – and are returned to them, and in which the whole process is powered by solar energy.'

p. 25 'takes energy' – Rifkin, J., and Howard, T. (1980), *Entropy – A New World View*, New York: The Viking Press.

p. 25 'a stable state' – Jackson, T. (1996), *Material Concerns – Pollution, profit, and quality of life*, London: Routledge, Ch. 2.

p. 27 'stood at just 0.8%' – Defra (2009), 'Household waste and

Recycling: 1983–4 to 2007–8'. Available at: http://www.Defra.
gov.uk/evidence/statistics/environment/waste/kf/wrkf04.htm

p. 27 'leapt from 7.5% in 1996/7 to nearly 38% in 2008/9' –
household waste recyling rates 1997/98–2007/08 are at:
http://www.defra.gov.uk/evidence/statistics/environment/
waste/kf/wrkf16.htm, and as of 2008/2009 the figure stands
at 37.6% as referenced in: Defra (2010) 'Municipal Waste
Statistics 2008/9.' Bulletin. Available at: *http://www.Defra.
gov.uk/evidence/statistics/environment/wastats/bulletin09.htm*

p. 27 'Austria and Germany' – European Commission (2010)
EuroStat: '40% of municipal waste recycled or composted in
2008' [press release]. Available at: http://epp.eurostat.ec.
europa.eu/cache/ITY_PUBLIC/8-19032010-AP/EN/8-19032010-
AP-EN.PDF [accessed 26 July 2010]. 'The overall best recycling
and composting performance in 2008 was found in Austria,
which achieved a 69% combined recycling and composting
rate for 2008. Other high performers were Germany (65%),
Belgium (60%) and the Netherlands (59%).'

p. 30 'if including water' – Biffaward (2006), *The Mass Balance
Movement: the definitive reference for resource flows within the
UK environmental economy*, London: Biffaward, 2006.
Available at: http://www.massbalance.org

p. 30 'in the next twenty years' – Friends of the Earth Europe,
Friends of the Earth Austria and ERI (2009), *Overconsumption
– Our use of the world's natural resources*. Available at:
http://www.foeeurope.org/publications/2009/Overconsumption
Sep09.pdf

p. 30 'less than a half' – WRAP (2010), personal communication.

Chapter 4

p. 34 'an endless cycle' – Steinhardt, P. J. and Turok, N. (2002),
'A Cyclic Model of the Universe', *Science*, 24 May, Vol. 296,
No. 5572, pp. 1436–49.

p. 34 '*Earthly Treasures*' – Petty, K., and Miazels, J. (2008), *Earthly
Treasures*, London: Eden Project Books, Transworld.

p. 35 'no less momentous' – see the excellent BBC4 series *Chemistry – a Volatile History* at http://www.youtube.com/watch?u=25lprEvoFJ8

p. 37 'biological activity' – Hazen, R. M., et al. (2008), 'Mineral evolution', *American Mineralogist*, 93: pp. 1693–1720.

p. 37 'up to AD 1800' – Ashby, M. (2009), *Materials and the Environment – Eco-Informed Material Choice*, USA: Butterworth-Heinemann.

p. 38 'stuff we have created' – Lloyd, C. (2008), *What on Earth Happened?*, London: Bloomsbury, Ch. 14.

p. 38 'settled "civilisations" to establish' – Headrick, D. (2009), *Technology: a World History*, OUP USA.

p. 38 'to throw clay pots' – Lloyd, C. (2008) *What on Earth Happened?* London: Bloomsbury, p. 126.

p. 39 'substitute for bronze' – Maurice, C., and Smithson, C. (1984), *The Doomsday Myth: 10,000 Years of Economic Crises*, USA: Hoover Institution Press. Quoted in Tilton, J. E. (2002), *On Borrowed Time? Assessing the Threat of Mineral Depletion*, RFF Press.

p. 39 'public baths and fountains' – Headrick, D. (2009), *Technology: a World History*, OUP USA.

p. 39 'continued to develop' – Gies, F., and Gies, J. (1995), *Cathedral, Forge, and Waterwheel: Technology and Invention in the Middle Ages*, reprinted edn, HarperCollins.

p. 39 'build houses and ships' – Williams, M. (2003), *Deforesting the Earth: from prehistory to global crisis*, University of Chicago Press.

p. 40 'kick-started the Industrial Revolution' – Nef, J. U. (1977), 'An Early Energy Crisis and Its Consequences', *Scientific American* 237, November, p. 140. This is the so-called Nef thesis – John Ulrich Nef posited that, in part, the Industrial Revolution was triggered by a sixteenth-century shortage of wood, which triggered demand for substitutes.

p. 40 'civic buildings of the nineteenth century' – Ashby, M. (2009), *Materials and the Environment – Eco-Informed Material Choice*, USA: Butterworth-Heinemann.

p. 40 'Pennsylvania in 1859' – Yergin, D. (1991), *The Prize: The Epic Quest for Oil, Money and Power*, New York: Simon & Schuster.

p. 40 'nearly a decade earlier' – Mir-Babayev, M. F. (2002), 'Azerbaijan's Oil History: A Chronology Leading Up to the Soviet Era', *Azerbaijan International Magazine*, Sherman Oaks, CA (US), AI 10.2 (Summer), pp. 34–41.

p. 41 'the coming of plastics' – Ashby, M. (2009), *Materials and the Environment – Eco-Informed Material Choice*, USA: Butterworth-Heinemann.

p. 41 'unique as far as we know' – Anders, B. (1968), *Earthrise* [photograph]. Available at: http://digitaljournalist.org/issue0309/lm11.html

Chapter 5

p. 42 'Sierra Club' – Mancell, P., and Mann, Co. *Material World*, Sierra Club Books, USA.

p. 44 'others found it cleansing' – Landy, M. (2008), *Everything must go!*, London: Ridinghouse.

p. 44 'populated his life' – Landy, M. (2010), in conversation with the author, 26 June.

Chapter 6

p. 47 'either grown or mined' – Eden Project (2001), mining exhibit.

p. 47 'regulate our blood sugar' – Whitney, E. R., and Rolfes, S. (2007), *Understanding Nutrition*, 11th edn, UK: Brooks and Cole.

p. 48 'including food production' – Ashby, M. (2009), *Materials and the Environment – Eco-Informed Material Choice*, USA: Butterworth-Heinemann.

p. 48 'concrete are produced a year' – Ashby, M. (2009),

Materials and the Environment – Eco-Informed Material Choice.
USA: Butterworth-Heinemann, Fig. 2.1, p. 17.

p. 49 '4 per cent annually' – Freedonia (2010), 'World Cement
to 2013 – Demand and Sales Forecasts, Market Share,
Market Size, Market Leaders, Company Profiles, Industry
Trends', press release. Available at: http://www.freedonia
group.com/World-Cement.html See also 'A Blueprint for
a Climate Friendly Cement Industry' at http://assets.
panda.org/downloads/englishsummary_lr_pdf.pdf, page 7
figure 1a.

p. 49 'concrete-making process' – WRAP (2004) 'The quality
protocol for the production of aggregates from inert waste'.
Available at: http://www.wrap.org.uk/downloads/0083_
Quality Protocol_A4.d9ba2d77.87.pdf [accessed 27 July 2010].

p. 50 'took another 10,000 years' – Macfarlane, A., and Martin, G.
(2002), *Glass: A World History*, University of Chicago Press, p. 10.

p. 52 'probably 4,000–5,000 years' – Macfarlane, A., and Martin, G.
(2002), *Glass: A World History*, University of Chicago Press, p. xi.

p. 52 'televisions and computer screens', see http://www.brit
glass.org.uk/Files/Recycling/GlassRecCounts.pdf, page 5.

p. 53 '32,000 wine bottles around the world' – WRAP (2004),
Recycled Glass Market Study and Standards Review, WRAP
online. Available at: http://www.wrap.org.uk/downloads/
GlassMktStudy2004.46ac408d.432.pdf [accessed 27 June 2010].

p. 53 'glass used in packaging' – http://www.defra.gov.uk/
environment/waste/producer/packaging/data.htm

p. 53 'majority of the rest' – UK Glass Manufacture (2008), A
Mass Balance Study. Available at: http://www.britglass.
org.uk/Files/Mass_Balance_Final.pdf [accessed on 27 July
2010] p.3.

p. 53 'for every 1,500 people' – Cocking, R. (2003), 'Glass
Recycling in the UK', Sustainable Waste Management
and Recycling: proceedings of the International
Symposium held at the University of Dundee, September
2003, p. 75.

p. 54 'uses 15% less energy' – WRAP (2004), Recycled Glass
Market Study and Standards Review, WRAP online.

Available at: http://www.wrap.org.uk/downloads/
GlassMktStudy2004.46ac408d.432.pdf [accessed 27 June
2010].

p. 54 'a recently developed habit' – British Glass (2008), 'The
Challenge Ahead – Glass collections', British Glass home
page. Available at: http://www.britglass.org.uk/Files/
Recycling/the_Challenge_FINAL.pdf

p. 54 'cement and other minerals' – BGS (2004), 'The
Economic Importance of Minerals in the UK', British
Geological Survey Commissioned Report. Available at:
www.bgs.ac.uk/downloads/start.cfm?id=1301

p. 55 'some form or other' – MII (undated), 'Common Minerals
and their Uses', Mineral Information Institute home page.
Available at: http://www.mii.org/commonminerals.html

p. 55 'metal for body piercings' – Gray, T. (2009), The Elements:
A visual exploration of every known atom in the universe, New
York: Black Dog and Leventhal Publishers, Inc., pp. 98–9
and 102–3.

p. 56 'employed in this way in developing countries' – ICMM
(2010), 'Artisinal and Small-scale Mining', International
Council on Mining & Metals home page. Available at:
http://www.icmm.com/page/37364/artisanal-and-small-scale-
mining

p. 56 'trained to use them' – Blacksmith Institute and Green
Cross Switzerland (2010), 'The 2009 World's Worst Polluted
Places: 12 Cases of Cleanup and Success'. Available at:
http://www.worstpolluted.org/

p. 57 'contaminate soil and water' – IIED Project and WBCSD
(2002), 'Breaking New Ground' Mining Minerals and
sustainable Development. Available at: http://www.wbcsd.
org/DocRoot/ev8jEJvTiMYd4mJhGGHO/finalmmsdreport.pdf

p. 57 'which killed 269 people' – Davies, M, et al. (2002), 'Mine
tailings Dams: when things go wrong', AGRA Earth &
Environmental Ltd. Available at: http://www.infomine.com/
publications/docs/Davies2002d.pdf

p. 57 'forest to mineral exploitation' – Campari, J. S. (2005), The
Economics of Deforestation: Dispelling the myth, London:

Edward Elgar Publishing Ltd, illustrated edn, p. 89. And Hall, A. L. (1989), *Developing Amazonia: Deforestation and Social Conflict in Brazil's Carajas Programme*, Manchester University Press.

p. 57 '101 things to do with a hole in the ground' – Pearman, G. (2009), *101 things to do with a hole in the ground*, Post-Mining Alliance and the Eden Project, Cornwall.

p. 57 'companies have all signed up to' – International Council on Mining and Metals (2003), Sustainable Development Principles. Available at: http://www.icmm. com/our-work/sustainable-development-framework/ 10-principles

p. 58 'restoration responsibilities' – Diamond, J. (2005), *Collapse – How Societies Choose to Fail or Survive*, London: Penguin. For a discussion on mining and mining companies, particularly in relation to Montana, US.

p. 58 'about 95% less' – http://www.alupro.org.uk/facts-and-figures.html

p. 66 'city of a million people' – Pearce, F. (2008), *Confessions of an Eco Sinner: Travels to Find Where My Stuff Comes From*, Eden Project Books, p. 191.

p. 59 '90,000 tonnes' – http://www.alupro.org.uk/facts-and-figures.html

p. 59 'EU every year' – Coca-Cola Company (2008), 'Coca-Cola enterprises leads transformation of drinks can design' [company press release, 22 September]. Available at: http://presscentre.coca-cola.co.uk/viewnews/coca cola enterprises leads transformation of drinks can design packaging.

p. 60 '42% of aluminium, overall' – Alupro (2009), 'Facts and Figures – All about aluminium packaging recycling in the UK. Available at: http://www.alupro.org.uk/facts-and-figures.html

p. 60 'in an English churchyard' – Pakenham, T. (1996), *Meeting with Remarkable Trees*, London: Weidenfield & Nicholson, p. 99. For example, the Much Marcle yew.

p. 61 'no less than twenty-six types' – UNEP (2000), 'Global

Distribution of Current Forests', United Nations Environmental Program, World Conservation Monitoring Centre. Available online: http://www.unep-wcmc.org/forest/global_map.htm

p. 61 'gone by the eleventh century' – Williams, M (2003), *Deforesting the Earth: from prehistory to global crisis*, University of Chicago Press. Or: James, N. D. G. (1981), A *History of English Forestry*, London: Blackwell.

p. 62 'production in tropical forests' – Forest Footprint Disclosure (2010), 'The Business Case by Commodity – Timber' [FFD home page]. Available at: http://www.forest disclosure.com/page.asp?p=4721. Seneca Creek Associates and Wood Resources International (2004) '"Illegal" Logging and Global Wood Markets: The Competitive Impacts on the U.S. Wood Products Industry', Maryland, USA.

p. 62 'wood used in the UK' – http://www.forestry.gov.uk/website/forstats2009.nsf/0/E9A1813CCFE0C9CA802575F50049AAF9

p. 63 'fairly energy-demanding' – Rivela, B., Moreira, M. T. and Feijoo, G. (2007), 'Life Cycle Inventory of Medium Density Fibreboard', *International Journal of Life Cycle Assessment*, Vol. 12, Iss. B, pp. 143–50.

p. 64 'life expectancy of five hundred years' – BBC (1999), 'Goat skin tradition wins the day', BBC Online: UK Politics. Available at: http://news.bbc.co.uk/1/hi/uk_politics/502342.stm [accessed 27 July 2010].

p. 65 'every tonne of paper' – Voith, M. (2010), 'The Paper Chase', *Chemical & Engineering News*, Vol. 88, No. 16, pp. 13–17.

p. 65 'Canada, USA and China' – Lewington, A. (2003), *Plants for People*, London: Eden Project Books, Transworld, p. 272.

p. 66 'completely illegally' – Forest Footprint Disclosure (2010), 'The Business Case by Commodity – Timber' [FFD home page]. Available at: http://www.forestdisclosure.com/page.asp?p=4721. Also Greenpeace (2010), 'How Sinar Mas is pulping the planet', Greenpeace International. Available at: http://www.greenpeace.org/international/en/publications/reports/SinarMas-APP/

p. 66 'Paper Trails' – Haggith, M. (2008), *Paper Trails*, London: Virgin Books.

p. 66 'levels by 2020' – Forest Footprint Disclosure (2010), 'The Business Case by Commodity – Timber' [FFD home page]. Available at: http://www.forestdisclosure.com/page.asp?p=4721.

p. 66 'newspapers and magazines' – CPI (2009), Confederation of Paper Industries: Industry Facts. Available at: http://www.paper.org.uk/information/statistics/2009Industry Facts.pdf

p. 66 'consumer of paper in the world' – 14th per Capita as of 2005, 5th overall for 2005, World Resources Institute (2010), 'Resource Consumption: Paper and paperboard consumption. Units: Metric tons', Searchable EarthTrends Database. Available at: http://earthtrends.wri.org.uk

p. 66 'reach 87% of the adult population' – PPA (2010), 'Key facts and figures', Periodical Publishers Association home page. Available at: http://www.ppa.co.uk/ppa-marketing/why-magazines/

p. 66 'missing a sale' – Balanced Reading – Mass Balance of the UK Magazine Publishing Sector, Biffaward 2005. Also for the same phenomenon in the US: The Paper Project: Co-op America, Independent Press Association, Conservatree (2001), 'Turning the page: Environmental Impacts of the Magazine Industry: Recommendations for improvement.' Available at: http://www.greenamericatoday.org/pdf/whitepapermagazines.pdf

p. 67 '12–13 million tonnes' – WRAP (2010), 'Market Situation Report 2009/2010 Realising the Value of Recovered Paper: An Update'. Available at http://www.wrap.org.uk/down loads/WRAP_Paper_market_situation_report_Feb2010.230cc0 ac.8440.pdf

p. 67 'copies of this book' – Assuming that a copy weighs around 500g.

p. 67 'makes a very good fuel' – UPM (2009), 'Wood fibres in the paper cycle'. Available at: http://w3.upm-kymmene. com/upm/internet/cms/upmmma.nsf/lupgraphics/wood_ fibers_inthe_paper_cycle.pdf/$file/wood_fibers_inthe_paper_cycle.pdf

p. 67 'on a single site' – PWC (2010), 'Viewpoints of CEOs in the forest, paper & packaging industry worldwide 2010 Edition', PricewaterhouseCooper CEO Perspectives. Available at: http://www.pwc.com/qx/en/forest-paper-packaging/ceo2009

p. 68 'on average 80% recycled' – DEFRA (2009), 'Voluntary agreements for newspapers, direct mail and magazines', Department for Environment, Food and Rural Affairs; Environment; Producer Responsibility. Available at: http://www.Defra.gov.uk/environment/waste/producer/voluntary/index.htm

p. 68 'using 100% recycled stock' – Aylesford Newsprint (2010), 'The History of Aylesford Newsprint'. Available at: http://www.aylesford-newsprint.co.uk/Students-HistoryAylesford.asp

p. 68 'around a third' http://www.Defra.gov.uk/evidence/statistics/environment/waste/download/xis/wrtb20.xls
The domestic recycling rate for paper and board is given as 33.4% for 2007.

p. 68 'nearly 70%' – WRAP (2010), 'Market Situation Report 2009/2010 Realising the Value of Recovered Paper: An Update'. Available at: http://www.wrap.org.uk/dowloads/WRAP_Paper_market_situation_report_Feb2010.38603c6f.8440.pdf Cites 67% for 2008.

p. 68 'back into the recycling process' – Voith, M. (2010), 'The Paper Chase', Chemical & Engineering News, Vol. 88, No. 16, pp. 13–17. 'The decline of print newspapers and magazines is a source of concern for journalists, but it also has an impact on the worldwide paper market. After coming to depend on readers to dutifully recycle their old periodicals, paper manufacturers are now finding that supplies of high-quality paper for recycling are becoming scarce. Used white office paper is also getting increasingly hard to find – and much more costly – thanks to growth in electronic communications. High-quality white paper is desired as a raw material for new paper. It contains long cellulose fibers with intact cell walls that can be used to make high-margin products for printing, packaging, or household use.'

p. 68 'landfill every year' – WRAP (2010), 'Market Situation Report 2009/2010 Realising the Value of Recovered Paper: An Update'. Available at: http://www.wrap.org.uk/dow loads/WRAP_Paper_market_situation_report_Feb2010.38603c 6f.8440.pdf Some 4 million tonnes not recycled in 2008 – exact amount to landfill not known, but probably at least three-quarters.

p. 70 'fairly difficult chemistry' – Science Museum (2009), '100 Years of Making Plastics', Science Museum home page. Available at: http://www.sciencemuseum.org.uk/visit museum/galleries/plasticity.aspx

p. 70 'Bakelite Museum' – Ghent Virtual Bakelite Museum (2007). Available at: http://juliensart.be/bakeliet/english.html

p. 70 'chlorine in salt' – Bellis, M. (20??), 'The History of Plastics: the first man-made plastics – Parkesine,' About.com. Available at: http://inventors.about.com/od/ pstartinventions/a/plastics.htm

p. 71 'consumed every year' – BPF (2009), 'About the British Plastics Industry', British Plastics Federation. Available at: http://www.bpf.co.uk/industry/Defoult.aspx

p. 71 '245 million tonnes in 2008' – Plastics Europe (2009), 'The Compelling Facts About Plastics 2009' An analysis of European plastics production, demand and recovery for 2008'. Available at: http://www.plasticseurope.org/plastics-industry/market-data.aspx

p. 71 'apart from steel' – Ashby, M. (2009), Materials and the Environment – Eco-informed Material Choice, USA: Butterworth-Heinemann.

p. 72 'In Europe' – BPF (2009), 'About the British Plastics Industry', British Plastics Federation. Available at: http://www.bpf.co.uk/industry/Default.aspx

p. 72 'used every year' – BPF (2009), 'Recycling', British Plastics Federation. Available at: www.bpf.co.uk/sustainability/ plastics_recycling.aspx

p. 74 'designed out' – http://www.green-alliance.org.uk/waste-content.aspx?id=4276 See presentation by Chris Dow in the slide set.

p. 75 'supplanted along the way' – Edwards, R., and Kellet, R. (2000), *Life in Plastic – the Impact of plastics on India*. Other India Press.

p. 76 'at least 50,000 years' – Lloyd, C. (2008), *What on Earth Happened?*, London: Bloomsbury, p. 102.

p. 76 'pebbles and teeth' – Schoeser, M. (2003), *World Textiles – A Concise History*, London: Thames & Hudson, p. 10.

p. 76 'depending on the part of the world' – Schoeser, M. (2003), *World Textiles – A Concise History*, London: Thames & Hudson, p. 15.

p. 76 'roots and bark' – Schoeser, M. (2003), *World Textiles – A Concise History*, London: Thames & Hudson, p. 28.

p. 77 'earliest was nylon' – Handley, S. (1999), *Nylon: the Story of a Fashion Revolution*, The Johns Hopkins University Press.

p. 77 '2,700 litres per T-shirt' – WWF (2003), 'Thirsty Crops: our food and clothes: eating up nature and wearing out the environment', Living Waters Publication, p. 5. Available at: http://assets.panda.org/downloads/wwfbookletthirsty crops.pdf

p. 77 'global insecticide use' – Maxwell, D., for DEFRA (2007) Sustainable Clothing Roadmap, Briefing Note, December 2007: Sustainability impacts of clothing and current interventions, GVSS for Defra. Available at: http://www.defra. gov.uk/environment/business/products/roadmaps/clothing/ index.htm

p. 77 'emissions to local pollution' – Turley, D., et al. (2009), 'The role and business case for existing and emerging fibres in sustainable clothing'. Final report to the Department for Environment, Food and Rural Affairs, London, July. Available at: http://www.Defra.gov.uk/environ-ment/business/scp/evidence/theme2/products0809.htm

p. 78 'expert, and therefore expensive, sorting' – BIR (2009), 'Textiles: Materials', Bureau of International Recycling. Available at: http://www.bir.org/industry/textiles. Textiles have to be sorted into different grades (i.e. whether sufficient quality for reuse) and also into the different material type.

p. 78 'generated each year' – DEFRA (2009), 'Maximising Reuse and Recycling of UK Clothing and Textiles' [report summary]. Available at: http://randd.defra.gov.uk/Default.aspx?Menu=Menu&Module=More&Location=None&Completeed=1&ProjectID=16096

p. 78 'worn at all' – http://www.churchill.com/press-Releases/06012006.htm

p. 79 'hours or days' – DEFRA (2009), 'Maximising Reuse and Recycling of UK Clothing and Textiles' [report summary]. Available at: http://randd.defra.gov.uk/Default.aspx?Menu=Menu&Module=More&Location=None&Completeed=1&ProjectID=16096

p. 79 '1839 by Charles Goodyear' – Goodyear (2009), 'History: The Charles Goodyear Story'. Available at: http://www.goodyear.com/corporate/history/history_story.html

p. 79 'smouldered for thirteen years' – Rowe, M. (2002), 'Waste tyres' environmental impact', Society, Guardian. Available at: http://www.guardian.co.uk/society/2002/may/15/environment.waste [accessed 27 July 2010].

p. 79 'work by WRAP' – http://www.wrap.org.uk/downloads/Tyres_FAQ_Report.2758cdbd.4111.pdf

p. 80 'nearly impossible to recycle' – Ashby, M. (2009), Materials and the Environment – Eco-Informed Material Choice, USA: Butterworth-Heinemann.

p. 80 'reclaim the materials' – Albers, H. (2007), 'Managing long-term environmental aspects of wind turbines: a prospective case study', International Journal of Technology, Policy, and Management, Vol. 7, No. 4.

p. 81 'freecycle website' – www.freecycle.org

p. 82 'wastefulness as well' – Peters, G. (2008), 'From production-based to consumption-based national emissions inventories', Ecological Economics, Iss. 65, pp. 13–23.

Chapter 7

p. 87 'basic life-support systems' – Steffen, W., Crutzen, P. J., and McNeill, J. R. (2007), 'The Anthropocene: Are humans now overwhelming the great forces of nature?', *Ambio*, December, Vol. 36, Iss. 8, pp. 614–21.

p. 89 'amount on the surface' – Barlow, M., and Clarke, T. (2002), *Blue Gold: The Battle Against Corporate Theft of the World's Water*, USA: McClelland & Stewart.

p. 91 'by the 2010 earthquake' – Than, K. (2010), 'Haiti Earthquake, Deforestation heightens landslide risk', *National Geographic*. Available at: http://news.national geographic.co.uk/news/2010/01/100114-haiti-earthquake-landslides

p. 91 'degradation including deforestation' – Huffington Post (2010), 'Beijing Sandstorm 2010: Photos of major China storm', Huffington Post, Associated Press Photos. Available at: http://www.huffingtonpost.com/2010/03/20/beijing-sandstorm-2010-ph_n_506875.html

p. 92 'and, of course, mammoths' – Bryson, B. (2003), *A Short History of Nearly Everything*, London: Doubleday.

p. 92 '1,000 times that level' – Baillie et al. (2004), *IUCN Red List of Threatened Species: A Global Species Assessment*, IUCN, Gland.

p. 92 'every twenty minutes' – Conservation International (2010), 'How do we set our Clock?' Available at: http://www.conservation.org/act/get_involved/Pages/stop-the-clock-methodology.aspx

p. 92 'human beings for the job' – Bryson, B. (2003), *A Short History of Nearly Everything*, London: Doubleday, p. 572.

p. 92 'friend Stephen Fry' – Carwardine, M. (2009), *Last Chance to See – in the footsteps of Douglas Adams*, London, Harper Collins.

p. 93 'and they are many' – Green Facts (2006), 'Biodiversity, ecosystem functioning, ecosystem services, and drivers of change', Greenfacts.org. Available at: http://www.green facts.org/en/global-biodiversity-outlook/figtableboxes/figure-1-1.htm

p. 94 'last three decades' – Brown, L. R. (2009), *Plan B 4.0: Mobilizing to Save Civilization*, Washington DC: Earth Policy Institute, pp. 5–6. Also McNeill, J. R. (2001), *Something New Under the Sun: an environmental history of the twentieth century*, USA: Penguin, Ch. 2.

p. 95 'drought and flood in turn' – Randerson, J. (2010), 'IPCC denies newspaper claim that it overstated costs of natural disasters', *Guardian*, 26 January. Available at: http://www.guardian.co.uk/environment/2010/jan/26/ipcc-natural-disasters-climate-change [accessed 28 July 2010].

p. 95 'doubling every twenty years' – Global water use doubled between 1940 and 1980, and it is expected to double again by 2000 – UN (1987), 'Report of the World Commission on Environment and Development: Our Common Future', Ch. 11. Available at: http://www.un-documents.net/wced-ocf.htm. More usually known as the Brundtland Commission.

p. 95 'per person is rising' – Barlow, M., and Clarke, T. (2002), *Blue Gold: The Battle Against Corporate Theft of the World's Water*, USA: McClelland & Stewart.

p. 95 'a regional one' – Meadows, D., et al. (2004), *Limits to Growth: The 30-Year Update*, Chelsea Green Publishing Company and Earthscan.

p. 96 'pumped up from deep underground' – Kamel, S., and Dahl, C. (2005), 'The economics of hybrid power systems for sustainable desert agriculture in Egypt', *Energy*, Vol. 30, Iss. 8.

p. 96 'very unequal' – Black, M. (2004), *The No-Nonsense guide to Water*, Oxford, New Internationalist Publications.

p. 96 'over eight in agriculture' – Hoekstra et al. (2009), 'Water Footprint Manual: State of the Art', Water Footprint Network. Available at: http://www.waterfootprint.org/downloads/WaterFootprintManual2009.pdf

p. 96 'the clothes we wear' – WWF (2003), 'Thirsty Crops: our food and clothes: eating up nature and wearing out the environment'. Living Waters Publication, p. 5. Available at: http://assets.panda.org/downloads/wwfbookletthirstcrops.pdf

p. 97 'plastic a massive 185 litres' – Treloar, G., et al. (2004), 'Embodied Water of Construction', *Environment Design Guide*, Royal Australian Institution of Architects.

p. 97 'global agriculture' – Chapagain, A.K. & Hoekstra. A.Y (2004) – 'The Water Footprints of Nations – Volume 1: Main Report', UNESCO-IHE.

p. 97 'for processing the ores' – Technomine (2007), 'Australian mine water use' [news story]. Available at: http://technology.infomine.com/articles/1/737/mining.water.australia/australian.mine.water.aspx

p. 98 'near-desert areas' – NASA (2009), 'Aral Sea Continues to Shrink', *Earth Observatory*. Available at: http://earthobservatory.nasa.gov/IOTD/view.php?id=39944

p. 98 'repairing the damage' – Pearce, F. (2006), *When the Rivers Run Dry: Water – The Defining Crisis of the Twenty-First Century*, London: Eden Project Books, Transworld.

p. 98 'collapse of civilisations' – Diamond, J. (2005), *Collapse – How Societies Choose to Fail or Survive*, London: Penguin.

p. 99 'implications of climate change' – News Distribution Service, 10 March 2010: 'Health to be at the centre of the fight against climate change'.

p. 101 'very uncertain territory' – Lynas, M. (2007), *Six Degrees: Our Future on a Hotter Planet*, London: Fourth Estate.

p. 102 'the Thames every year' – Brown, P. (2004), '£2bn tunnel to carry sewage under Thames', *Guardian*. Available at: http://www.quardian.co.uk/environment/2004/apr/10/uknews.pollution [accessed 14 July 2010].

p. 103 'concentrations of multiple chemicals' – Grandjean, P., and Landrigan, P. J. (2006), 'Developmental neurotoxicity of industrial chemicals', *Lancet*, 368(9553): pp. 2167–78.

p. 104 'elements in our bodies' – Whitney, E. R., and Rolfes, S. (2007), *Understanding Nutrition*, 11th edn, UK: Brooks and Cole.

p. 104 'twenty times above safe levels' – Montana Bureau of Mines and Geology (2004), 'Berkeley Pit and BMF Operable Unit', Environmental Assessment [home page]. Available at: http://www.mbmg.mtech.edu/env/env-berkeley.asp

p. 105 'local drinking water' – Diamond, J. (2005), *Collapse – How Societies Choose to Fail or Survive*, London: Penguin.

p. 105 'into people's homes' – LeCain, T. J. (2009), *Mass Destruction: the men and giant mines that wired America and scarred the planet*, USA: Rutgers University Press.

p. 105 'traces of metals' – Barlow, M., and Clarke, T. (2002), *Blue Gold: The Battle Against Corporate Theft of the World's Water*, USA: McClelland & Stewart.

p. 107 'UV has been avoided' – Slaper, H., et al. (1996), 'Estimates of ozone depletion and skin cancer incidence to examine the Vienna Convention achievements', *Nature*, Iss. 384(6606), pp. 256–8.

Chapter 8

p. 109 'giant hamster' – New Economics Foundation (nef), 2010 'Growth Isn't Possible'. Available at: http://www.new economics.org/publications/growth-isnt-possible

p. 109 'businessmen, statesmen and scientists' – Meadows, D., et al. (2004), *Limits to Growth: The 30-Year Update*, Chelsea Green Publishing Company and Earthscan, p. 338.

p. 111 'less than 1% now' – Technology Strategy Board (2009), 'Resource Efficiency Strategy 2009–2012', p. 12. Available at: http://www.innovateuk.org/_assets/pdf/Corporate-Publi cations/ResourceEfficiencyStrategyT09_080.pdf

p. 111 'a staggering 500,000 tonnes' – Hall, T. (2008), 'Abiotic rucksack of various raw materials', Table 4.2, *Use of Composite Environmental Indicators in Residential Construction*, USA: MIT Press.

p. 111 'a source as the ore' – Farago, M., et al. (1996), 'Platinum metal concentrations in urban road dust and soil in the United Kingdom', *Fresenius' Journal of Analytical Chemistry*, 354(5), 660–63. And Mungall, J. E., & Naldrett, A. J. (2008), 'Ore Deposits of the Platinum-Group Elements', *ELEMENTS*, Vol. 4, pp. 253–8.

p. 112 'high environmental impact to extract' – Oakdene

Hollins (2008), 'Material Security – Ensuring resource availability for the UK economy', strategic report produced by
the Resource Efficiency Knowledge Transfer Network.
Available at: www.oakdenehollins.co.uk/pdf/material_
security.pdf

p. 113 'controlled by China' – Oakdene Hollins (2008), 'Material
Security – Ensuring resource availability for the UK
economy', strategic report produced by the Resource
Efficiency Knowledge Transfer Network. Available at:
www.oakdenehollins.co.uk/pdf/material_security.pdf

p. 113 'future riches' – *Money Week*, Issue 497, http://www.
moneyweek.com/investments/commodities/investing-in-
rare-metal-stocks-mining-rare-metals-49726.aspx

p. 113 'much more abundant' – Lahiri, A., and Jha, A. (2009),
'Selective separation of rare earths and impurities from
ilmenite ore by addition of K+ and Al3+ ions',
Hydrometallurgy, Iss. 95, pp. 254–61.

p. 114 'little more than a year' – http://www.imf.org/external/
pubs/ft/weo/2009/update/02/index.htm

p. 114 'economy continue to rise' – Behrens, A., et al. (2007),
'The material basis of the global economy: Worldwide
patterns of natural resource extraction and their implications for sustainable resource use policies', *Ecological
Economics* 64 (2), pp. 444–53.

p. 114 'hurtle down that path' – Meadows, D., et al. (2004),
Limits to Growth: The 30-Year Update, Chelsea Green
Publishing Company and Earthscan, p. 51. 'Our concern
about collapse does not come from belief that the world is
about to exhaust the planet's stocks of energy and raw.
Every scenario produced by World3 shows that the world
in 2100 still has a significant fraction of the resources that
it had in 1900. In analysing World3 projections our concern
rather arises from the growing cost of exploiting the
globe's resources and sinks. Data on these costs are inadequate, and there is substantial debate on the issue. But we
conclude from the evidence that growth in the harvest of
renewable resources, depletion of non renewable resources,

and the filling of the sinks are combining slowly and inexorably to raise the amount of energy and capital required to sustain the quantity and the quality of material flows required by the economy ... Eventually they will be high enough that growth in industry can no longer be sustained. When that happens the positive feedback loop that produced expansion in the material economy will reverse direction; the economy will begin to contract.'

p. 115 'regenerative capacity around 1980' – Wackernagel, M., et al. (2002), 'Tracking the Ecological Overshoot of the Human Economy', *Proceedings of the National Academy of Sciences*, Vol. 99, No. 14, pp. 9, 266–71.

p. 115 'by nearly 30 percent' – Brown, L. R. (2009), *Plan B 4.0: Mobilizing to Save Civilization*, Earth Policy Institute, Washington DC. 'Paul Hawken, author of *Blessed Unrest* (2007), puts it well: "At present we are stealing the future, selling it in the present, and calling it gross domestic product"', p. 15.

p. 116 'carry on forever' - Diamond, J. (2005) *Collapse – How Societies Choose to Fail or Survive*. Penguin, England. For a discussion on mining and mining companies, particularly in relation to Montana, US.

p. 117 'to greater consumption' - Polimeni, J.M. et al (2008), *The Jevons Paradox and the Myth of Resource Efficiency Improvement*, London, Earthscan Books.

p. 118 '4.6 to 10.3 tonnes per year' – Behrens, A., et al. (2007), 'The material basis of the global economy: Worldwide patterns of natural resource extraction and their implications for sustainable resource use policies', *Ecological Economics*, Iss. 64, pp. 444–53.

p. 119 'concerned with environmental resilience' – Rockström, J. (2009), 'Planetary boundaries: exploring the safe operating space for humanity', *Ecology and Society*, Iss. 14, p. 32. Available at: http://www.ecologyandsociety.org/vol14/iss2/art32/

p. 120 'healthier than you think' – *New Scientist* (2010), 'Earth's

nine lives – why the planet is healthier than you think'
[front cover text], New Scientist, 24 February.

p. 120 'may contain crevasses' – Costanza, R. (1997), Frontiers
in Ecological Economics, Transdisciplinary Essays by Robert
Costanza, Cheltenham: Elgar, p. 14.

p. 120 'let nature choose them for us' – Meadows, D. (2007),
'The history and conclusions of the Limits to Growth',
System Dynamics Review, Vol. 23, No. 2, p. 194.

p. 121 'to get the same yield' – Benton, T., et al. (2010), 'Scale
matters: the impact of organic farming on biodiversity at
different spatial scales', Ecology Letters, Vol. 13, Iss. 7, pp.
858–69. 'The yield from organic fields was 55 per cent
lower than from conventional fields growing similar crops
in the same areas.'

p. 122 'at the expense of theirs' – nef (2007),
'Chinadependence: The second UK Interdependence
Report', New Economics Foundation. Available at: http://www.
neweconomics.org/sites/neweconomics.org/files/China
dependence_1.pdf

p. 122 'achieve efficiencies' – WRAP (2009), 'Meeting the UK
climate change challenge: The contribution of resource
efficiency'.

Chapter 9

p. 127 'giant stationery was not long-lasting' – Actually, it
became my pleasure to consume it, because Tracy mended
it and gave it to me as a present.

p. 129 'the Throwaway Age' – Packard, V. (1960), The Waste
Makers, USA: Pocket Books Inc., p. 7.

p. 130 'willing to take on the "affluent society"' – See also
Galbraith, J. K. (1958), The Affluent Society, Boston: Houghton
Mifflin.

p. 130 – 'off his chest' – Lawson, N. (2009), All Consuming,
London, Penguin Books.

p. 131 'enough economic growth' – Naish, J. (2008), Enough:

Breaking free from the world of excess, London: Hodder & Stoughton.

p. 132 'materialism becomes relative "is not clear cut"' – James, O. (2008), *The Selfish Capitalists – Origins of Affluenza*, London: Ebury Press.

p. 132 'mankind's most important assets' – de Botton, A. (2004), *Status Anxiety*. London: Hamish Hamilton.

p. 133 'conspicuous consumption' – 'Lunn, P. (2008), *Basic Instincts – Human Nature and the New Economics* London, Marshall Cavendish.

p. 133 'can get rather than what we need' – Girling, R. (2009), *Greed – Why We Can't Help Ourselves*, London: Transworld.

p. 134 'increasingly miserable' – see also Layard, R. (2005), *Happiness – Lessons from a New Science* London, Penguin Books.

p. 134 'on its head' - Miller, D. (2008), *The Comfort of Things* Cambridge, Polity Press; Miller D. (2010) *Stuff*, Cambridge, Polity Press.

p. 135 'on what "sustainable consumption" might mean' – Jackson, T., ed. (2006), *The Earthscan Reader in Sustainable Consumption*, Earthscan Reader Series.

p. 136 'as a teenager' – Le Guin, U.K. (1974) *The Dispossessed*, London, Gollancz S.F.

Chapter 10

p. 139 'at the root of our powerlessness' – Bolton, G. (2007), *Aid and Other Dirty Business*, London: Ebury Press, p. 310.

p. 140 'environmental policies profitable' – Diamond, J. (2005), *Collapse – How Societies Choose to Fail or Survive*, London: Penguin, p. 484.

p. 140 'aimed at the Green Consumer' – Elkington, J & Hailes, J (1988), *The Green Consumer Guide*, London, Victor Gollancz Ltd.

p. 140 'one group or another' – Clark, D. (2006), *The Rough Guide to Ethical Living*, Rough Guides, 1st edn.

p. 143 'aquatic life that can accompany eutrophication' – Macdonald, M. A., and Densham, J. M., et al. (2007), 'Too much of a good thing? The impacts of nutrients on birds and other diversity', in Hislop, Hannah, ed. (2007), *The Nutrient Cycle: Closing the Loop*, London: Green Alliance.

p. 143 'fertilisers and manures from agricultural land' – Hislop, Hannah ed. (2007), *The Nutrient Cycle: Closing the Loop*, London: Green Alliance, diagram, p. 12. Available at: http://www.green-alliance.org.uk/uploadedFiles/Publications/reports/TheNutrientCycle.pdf

p. 145 'sludge in agriculture' – European Commission Joint Research Centre Institute for Environment and Sustainability Soil and Waste Unit (2001), Organic Contaminants in Sewage Sludge for Agricultural use. Available at: http://ec.europa.eu/environment/waste/sludge/pdf/organics_in_sludge.pdf

p. 145 'gone in fifty years' time' – Steen, I. (1998), 'Phosphorus Recovery: Phosphorus Availability in the 21st Century: Management of a Non Renewable resource', *Phosphorus & Potassium*, Iss. 217.

p. 145 'protect supplies for their own agriculture' – *Fertilizer Week* (2008), 'Industry ponders the impact of China's trade policy', *Thursday Markets Report*, 24 April, British Sulphur Consultants, CRU.

p. 145 'only accounts for around 10%' – Wind, T., et al. (2007), 'The Role of Phosphate Detergents in the Phosphate-balance of European surface waters', *Official Publication of the European Water Association*.

p. 146 'According to Rose George' – George, R. (2008), *The Big Necessity – Adventures in the World of Human Waste*, London: Portobello Books, p. 52.

p. 146 'during the Second World War' – Joyce, J. (1922), *Ulysses*, Oxford University Press, p. 66.

p. 146 'Scott Brothers in Philadelphia' – Science Museum (2009), 'Brought to Life: Toilet paper, "Bronco" brand, London, England, 1935–50'. Image available at: http://www.science museum.org.uk/broughttolife/objects/display.aspx?id=1790

p. 146 'why on earth that should be' – Bog Standard (2009), 'Pupils' Site Area: Promoting better toilets for pupils'. Available at: http://www.bog-standard.org/pupils_history-aspx

p. 147 'fifty-seven sheets of toilet paper a day' – George, R. (2008), *The Big Necessity – Adventures in the World of Human Waste*, London: Portobello Books, p. 69.

p. 147 'dirtiest parts of their body' – George, R. (2008), *The Big Necessity – Adventures in the World of Human Waste*, London: Portobello Books, p. 53.

p. 147 '110 rolls a year' – Ethical Consumer (2007), 'Buyer's guide to toilet paper'. Available at: http://www.ethical consumer.org/FreeBuyersGuides/householdconsumables/Toiletpaper.aspx

p. 147 'super-soft kind increased 40% in 2008' – Walsh, B. (2009), 'A delicate undertaking', *Time*, 10 June. Available at: http://www.time.com/time/health/article/0,8599,1903778,00.html 'Toilet paper containing 100% recycled fiber makes up less than 2% of the U.S. market, while sales of three-ply luxury brands like Cottonelle Ultra and Charmin Ultra Soft shot up 40% in 2008. Compare the U.S. desire for an ever plusher flush with the more austere bathroom habits of Europe and Latin America, where recycled TP makes up about 20% of the at-home market.'

p. 147 'certain good management standards' – FSC (2009), 'What's the Forest Stewardship Council?' Available at: http://www.fsc-uk.org/wp-content/plugins/downloads-manager/upload/What%20is%20the%20Forest%20Stewardship%20Council%20web.pdf 'FSC certified forests must be managed to the highest environmental, social and economic standards. Trees that are harvested are replanted or allowed to regenerate naturally, but it does not stop there. The forests must also be managed with due respect for the environment, the wildlife and the people who live and work in them. This is what makes the FSC system unique and ensures that a forest is well-managed, from the protection of indigenous people's rights to the methods of felling trees.'

p. 147 'standards of certification such as PEFC' – PEFC is an international NGO aiming to promote Sustainable Forest Management through third-party certification – for more information see: http://www.pefc.co.uk/about-pefc/about

p. 148 '9% of world market pulp capacity' – FSC in email correspondence with the author, 9 July 2010.

p. 148 'but exclude "unacceptable" forestry' – FSC (2009) 'What's the Forest Stewardship Council?' Available at: http://www.fsc-uk.org/wp-content/plugins/downloads-manager/upload/What%20is%20the%20Forest%20Stewardship%20Council%20web.pdf

p. 149 'waste fibres from textile production' – Hickman, L (2009), 'Ask Leo: which toilet paper is the most ethical?' *Guardian*, 19 March. Available at: http://www.guardian.co.uk/environment/2009/mar/19/ethical-toilet-paper [accessed 28 July 2010].

p. 149 'rice, hemp, bamboo and wheat' – Robbins, N. (2010), 'Flushing Forests', World Watch, May/June. Available at: http://www.worldwatch.org/node/6403

p. 150 'taken from the roll' – Tree Hugger (2009), 'Squared toilet paper = less waste'. Available at: http://www.treehugger.com/files/2009/05/square-toilet-paper-shigeru-ban.php

p. 150 'twenty years of industrial emissions' – Moors for the Future (2007), 'Carbon Flux from the Peak District Moorlands', Research Note no. 12. Available at: http://www.moorsforthefuture.org.uk/mftf/downloads/publications/MFF_researchnote12_carbonflux.pdf

p. 150 'emissions from 75,000 households' – Act on CO_2 (2010) 'Dig with peat-free compost'. Available at: http://actonco2.direct.gov.uk/home/what-you-can-do/Out-shopping/Dig-with-peat-free-compost.html

p. 151 'relatively new use' – *Guardian* Online (2010), 'Compost and climate change: how they are related'. Available at: http://www.quardian.co.uk/peat-free-compost/climate-change [accessed 28 July 2010].

p. 151 'national park in Spain' – Tremlett, G. (2009), 'Spanish wetlands shrouded in smoke as climate change dries out peat', *Guardian*. Available at: http://www.guardian.co.uk/environment/2009/oct/19/las-tablas-water-wetlands-burn [accessed 28 July 2010].

p. 151 'watch it on YouTube' – Aljazeera Videos (2009), 'Burning issue of Spain's wetlands'. Available at: http://www.youtube.com/watch?v=qsORi5_YJoE&feature=youtube_qdata [accessed 28 July 2010].

p. 151 'municipal green waste' – Farrell, M., and Jones, D. (2010), 'Food waste composting: Its use as a peat replacement', *Waste Management*, Iss. 30, p. 1495.

p. 151 'campaign by WRAP' – WRAP (2009), 'Guide to buying peat free compost'. Available at: http://www.wrap.org.uk/downloads/Buying-Guide.6513c6cc.5122.pdf

p. 151 'probably did contain peat' – National Trust (2010), 'Go Peat Free: Making Gardens Greener'. Available at: http://www.nationaltrust.org.uk/main/w-chl/w-places_collections/w-gardens-greenergardens/w-gardens-make gardensgreener/w-gardens-makegardensgreener-gopeat free.htm

p. 152 '72% in 2007' – DEFRA (2008), 'Monitoring of peat and alternative products for growing media and soil improvers in the UK 2007'. Available at: http://www.Defra.gov.uk/evidence/science/publications/documents/peatAlternatives 2007.pdf

p. 152 'producing individual packets' – Walkers (2009), 'Walkers Carbon Savings' [company fact sheet]. Available at: http://www.walkerscarbonfootprint.co.uk/walkers_carbon_trust.html

p. 153 'astonishing comparisons' – Berners-Lee, M. (2010) – *How Bad are Bananas? The Carbon Footprint of Everything*, London, Profile Books.

p. 154 'thousand of products on the market' – European Commission (2009), EU Ecolabel, Licence Applications [graph]. Available at: http://ec.europa.eu/environment/ecolabel/about_ecolabel/facts_figures/evo01.gif

p. 155 'green claims code' – DEFRA (2003), 'Green claims

guidance'. Available at: http://www.Defra.gov.uk/environment/business/marketing/glc/claims.htm

p. 155 'withdraw the ad' – ASA (2010), 'Adjudications', Advertising Standards Authority. Available at: http://www.asa.org.uk/Complaints-and-ASA-action/Adjudications.aspx

p. 155 'shamed on the ASA website' – ASA (2010), 'Environmentally friendly advertising claims on the rise', Advertising Standards Authority. Available at: http://www.asa.org.uk/Resource-Centre/Hot-Topics/Environmental-claims-on-the-rise.aspx

p. 155 'an "expert panel" early in 2010' – Which? (2010), 'Which? Green Claims Study: Many of you are sceptical about green products'. Available at: http://image.guardian.co.uk/sys-files/Environment/documents/2010/04/29/greenwashwhich.pdf

p. 156 'embarrassed or ashamed?' – Single Planet Living (2010), 'Roadmap to sustainable development'. Available at: http://www.singleplanetliving.com/about1.shtml

p. 157 'examine the "secret life of stuff"' – Ryan, J. C., and During, A. T. (1997), Stuff: The Secret Lives of Everyday Things, Northwest Environment Watch, Seattle, USA.

p. 158 'sobering story' – Leonard, A with Conrad, A. (2010), The Story of Stuff: New York: Free Press.

p. 158 'Eco Sinner' – Pearce, F. (2008), Confessions of an Eco Sinner: Travels to Find Where My Stuff Comes From, Eden Project Books.

p. 158 'author Daniel Goleman' – Goleman, D. (2009), Ecological Intelligence – Knowing the Hidden Impact of What We Buy, London: Allen Lane.

p. 159 'product "wikis" and blogs' – Appropedia seems to be the most popular (www.appropedia.org): the others he lists are Good Guide (www.goodguide.com), Climate Counts (www.climatecounts.org) and Nature and More (www.natureandmore.com).

p. 159 'Nokia 3G mobile phone' – Nokia Finland (2005) Integrated Product Policy Pilot Project Stage 1 Report.

Available at: http://ec.europa.eu/environment/ipp/pdf/
nokia_mobile_05_04.pdf

p. 160 'as is often the case' – Retrevo (2010), 'Is there any
hope for Green Gadgets,' Retrevo Pulse [press release].
Available at: http://www.retrevo.com/content/blog/2009/
03/are-ce-industry%2526%2523039%3Bs-green-efforts-trouble
%3F Note, no indication is given as to the size of the sample.

Chapter 11

p. 164 'local firms' – For more information about Shetland
Heat Energy and Power Limited, visit www.sheap-ltd.co.uk

p. 165 The Scottish government wastes by 2025, Zero Waste
proposals at http://www.scotland.gov.uk/Topics/
Environment/waste-and-pollution/Waste-1/wastestrategy

p. 166 'redundant offshore installations' – For more information about decommissioning in Shetland, visit http://
www.shetlanddecommissioning.com/

p. 166 'such a small community' – From page 7 of the 'Orkney
and Shetland Area Waste Plan'. Available at: http://
www.shetland.gov.uk/waste/documents/AreaWastePlan.pdf

p. 167 'believe this' – Hall K.D., Guo J., Dore M., Chow CC
(2009), The Progressive Increase of Food Waste in America
and its Environmental Impact. Available at: PLoS_ONE 4(11):
e7940.doi:10.1371/journal.pone.0007940

p. 169 'cups each year' – Dineen, Shauna (Nov. – Dec. 2005),
'The Throwaway Generation: 25 Billion Styrofoam Cups a
Year', E-The Environmental Magazine. Available at:
http://www.emagazine.com/view/?2933

p. 169 '58 Boeing 747s' – 'Airlines and Recycling: The Not-
So-Green Skies' *Scientific American Magazine*, September
2009.

p. 169 'in paper' – EPA (2008), 'Land quality: Quick facts and
figures'. Available at: http://www.epa.gov/osw/conserve/
materials/paper/faqs.htm

p. 169 'opening junk mail' – Swanson, R. (1995), *Margins:*

Restoring Emotional, Physical, Financial and Time Reserves to Overloaded Americans.

p. 170 'waste in the world' – OECD Environmental Data Compendium (2008).

p. 170 'developing over here' – Davis, S. J., & Caldeira, K. (2010). Consumption-based accounting of CO_2 emissions, Proceedings of the National Academy of Sciences, 107(12), 5687–5692. Available at: http://www.ciw.edu/news/carbon_emissions_outsourced_developing_countries

p. 173 'approximately 80% in Kamikatsu' – Green Alliance (2005), 'Kamikatsu, Japan' case study. Available at: http://www.green-alliance.org.uk/uploadedFiles/Our_Work/Kamikatsu.pdf

p. 174 'home processing of organic waste' – BBC (2005), 'World Without Waste'. Available at: http://news.bbc.co.uk/1/hi/programmes/documentary_archive/4499238.stm [accessed July 28 2010].

Chapter 12

p. 180 'reason the sky looks blue' – NASA (2005), 'Stardust: NASA's Comet Sample Return Mission: Aerogel' [website]. Available at: http://stardust.jpl.nasa.gov/tech/aerogel.html The comet debris hits the aerogel sail at six times the speed of a rifle bullet. The lack of density slows the dusts down, but prevents them heating up through excess friction – which means they do not alter their chemical make-up – so scientists can find out what comets' tails consist of, rather than examining the burned debris.

p. 181 'a note written on it' – Molotch, H. (2003), *Where Stuff Comes From*, New York: Routledge, p. 47.

p. 181 'around 90% fail' – McGeevor, K. (2009), 'To Choose or not to Choose', *Green Alliance Inside Track*, Iss. 23, p. 7.

Chapter 13

p. 188 'processing the Eden metal' – Paterson, M. (2007), 'Eden Project, from rock to roof – responsible sourcing at Eden'. Available at: http://www.edenproject.com/documents/rock-to-roof.pdf

p. 189 'the right recycling technology' – Recycling News Portal (2010), 'Missing out on potential income from recycling unused electronic equipment'. Available at: http://www.recyclingnewsportal.com/Recycling_article9753.html

p. 189 'who we'd like to be' – Miller, D. (2009), *Stuff*, Cambridge: Polity Press.

p. 190 'keep our homes warmer' – Geoghegan, T. (2009), 'What central heating has done for us', *BBC News Magazine*, 1 October. Available at: http://news.bbc.co.uk/1/hi/magazine/8283796.stm

p. 191 'food for available land' – Lee, M. (2007), *Eco-Chic – the savvy shopper's guide to ethical fashion*, London: Gaia Books.

p. 191 'influenced English design (as chintz)' – Selvedge (2008), *Selvedge*, Iss. July/August, p. 19.

p. 191 Dutch organisation 'Made-By' – http://www.made-by.nl/downloads/EnvironmentalBenchmarkFibersExternalParties2009.pdf

p. 191 'textile industy' – 'How should we measure?', *Ecotextile News*, (July 2010).

p. 192 detailed work on this area – Hoekstra et al. (2009), 'Water Footprint Manual: State of the Art', Water Footprint Network. Available at: http://www.waterfootprint.org/downloads/WaterFootprintManual2009.pdf

p. 193 'most water-efficient way' – Gerbens-Leenes, W., and Hoekstra, A., et al (2009), 'The water footprint of bioenergy', *Proceedings of the National Academy of Sciences*, Vol., 106, Iss. 25, pp. 10219–23.

p. 193 'as well as food and drink' – Segal, R, and MacMillan, T. (2009), 'Water labels on food: Issues and recommendations', *Food Ethics Council and Sustain*. Available at: http://www.sustainweb.org/pdf/water_labels_on_food.pdf

p. 194 'many other models' – Birtles, P. (1980), *Mosquito: A Pictorial History of the DH98*, London: Jane's Publishing Company Ltd.

p. 194 'world's supply of cement' – Hammond, E. (2009), 'Cement makers seek to spread risk', *Financial Times*, 30 December.

p. 194 'worsen coastal erosion' – Cemex (2010), 'Marine Aggregate Solutions: Environment' [home page]. Available at: http://www.cemex.co.uk/aa/aa_ma_en.asp

p. 194 'recycled glass' – WRAP (2008), Choosing construction products: Guide to the recycled content of mainstream construction products; also WRAP (2009), Compendium of Design Case Studies: Designing Out Waste Competition.

p. 195 'recyclability of the material' – Berg, W. T. (1993), 'Fiber Reinforced Concrete', *The Technical Advisor*, June. Available at: http://www.retailsource.com/information/fiber_rc/fiber_rc.html

p. 196 '12% of the country's energy' – FAO (1997), 'The role of wood energy in Europe and OECD: Wood Energy Today for Tomorrow', Regional Studies, Rome. Available at: http://www.fao.org/docrep/w7407e/w7407e00.htm

p. 196 'no more residue than unglued timber' – KLH (2010) KLH Massivholz Gmbh [website]. Available at: http://www.klh.at/unternehmen.html?L=3

p. 196 'buildings made from hemcrete' – TBAC (2009), 'Green Acres Courtyard, Chalgrove, Hemcrete Development'. Available at: http://www.tbaccentres.co.uk/greenacres/

p. 198 'hardbacked books for the first time' – Barnett, E. (2010), 'E-books outsell hardbacks on Amazon', Telegraph, 20 July. Available at: http://www.telegraph.co.uk/technology/news/7900191/E-books-outsell-hardbacks-on-Amazon.html [accessed 13 August 2010].

p. 198 'already serving the site' – Marten, P. (2009), 'Google Finds Itself in Finland', This is Finland [home page]. Available at: http://finland.fi/Public/default.aspx?contentid=163306&nodeid=41805&culture=en-US

p. 199 'more than a hundred years' – Simonite, T., and Le Page, M. (2010), 'Digital doomsday: the end of Knowledge',

New Scientist, 2 February, Iss. 2745. Available at:
http://www.newscientist.com/article/mg20527451.300-
digital-doomsday-the-end-of-knowledge.html

Chapter 14

p. 200 'economic system as possible' – One of the most
eloquent exponents is Murray, R. (2002), *Zero Waste*,
Greenpeace UK.

p. 201 'very forcefully' – Braungart, M. McDonough, W. (2002)
Cradle to Cradle – Remaking the way we make things, USA,
North Point Press.

p. 203 'phase called evolution' – Benyus, J. (1997), *Biomimicry –
Innovation Inspired by Nature*, London: HarperCollins. See
also work by Professor Julian Vincent, www.material
beliefs.com/collaboration/Julian-v.php

p. 203 'UK exponents of these ideas' – Pawlyn, M. (forthcoming), *Biomimicry in Architecture*, RIBA Publications.

p. 203 'at a time of our choosing' – Benyus, J. (1997), *Biomimicry
– Innovation Inspired by Nature*, London: HarperCollins.

p. 204 'plastic bags somewhere in the world' – See the excellent Resource Recovery Forum service for regular stories.

p. 204 'food supply' – de Braganca, Radek Messias, and Fowler,
Paul (2004), 'Industrial Markets for Starch', The
Biocomposite Centre, University of Wales, Bangor.

p. 205 'return to nature's cycles' – BASF (2010), 'Ecoflex
Biodegradable Plastic' [home page]. Certification documents
avaiiable at: http://www.plasticsportal.com/products/
ecoflex.html

p. 205 'at 20%' – NNFCC Market Review for European
Commission Lead Market Initiative 2009 (forthcoming).

p. 205 'replaced by the new materials' – Shen, L., et al. (2009),
'Product overview and market projection of emerging bio-
based plastics', *PRO-BIP 2009, Final report,* June, *European
Bioplastics,* Fig 1–1. Or: Patel, M. K., and Crank, M. (2007),
'Projections for the production of bulk volume bio-based

polymers in Europe and environmental implications',
Journal of Biobased Materials and Bioenergy, Iss. 1, pp. 437–53.

p. 206 'claimed to compost at home' – WRAP (2009), 'Home composting' [website]. Available at: http://www.wrap.org.uk/local_authorities/research_guidance/home_composting/home_compost.html Also Sloley, C. (2010), 'Subsidising composting bins "could save £600,000"', Let's Recycle, 20 July. Available. at: http://www.letsrecycle.com/do/ecco.py/view_item?listid=37&listcatid=5581&listitemid=55878 Also: Linda Crichton, WRAP, personal communication to the author: in 2007 36% of English households composted at home (figure will be slightly lower for the UK – about 35%).

p. 207 'interest began to wane' – Green Alliance (2007–10), 'Designing Out Waste' [website]. Available at: http://www.green-alliance.org.uk/waste-content.aspx?id=2340

p. 207 'promoting then thoroughly' – Green Alliance (2009), 'Designing Out Waste: Compostable packaging: where next?' [website]. Available at: http://www.green-alliance.org.uk/waste-content.aspx?id=2341

p. 207 'update meeting in 2009' – Green Alliance (2009), 'Designing Out Waste: Compostable packaging: where next?' [website]. Available at: http://www.green-alliance.org.uk/waste-content.aspx?id=2341

p. 209 'thus driving up prices' – Mitchell, D. (2008), 'A Note on Rising Food Prices', World Bank Development Economics Group [DEC] World Bank Policy, Research Working Paper No. 4682. Mitchell's estimates are at the high end at 70% of the price increases being attributable to biofuels, but other commentators agree with at least 30%.

p. 209 'had been taken into account' – Plevin R., et al. (2008), 'Uncertainty analysis of land-use change carbon release', EEA Expert Meeting, Copenhagen.

p. 209 'encroach onto wild lands' – RFA (2008), *The Gallagher Review of the indirect effect of biofuels production*, Renewable Fuels Agency, July. Available at: http://www.renewablefuelsagency.gov.uk/reportsandpublications/reviewoftheindirecteffectsofbiofuels

p. 209 'together in a mixed system' – Tilman, D., and Socolow, R., et al. (2009), 'Beneficial Biofuels – The Food, Energy, and Environment Trilemma'. *Science* 325, 270–1.

p. 209 'energy captured from the crop' – Buttazzoni, M. (2009), 'GHG Emission Reductions with Industrial Biotechnology: Assessing the Opportunities', WWF report. Available at: http://assets.panda.org/downloads/wwf_biotech_technical_report.pdf

Chapter 15

p. 211 'fail to eat some 25%' – WRAP (2009), 'Household Food and Drink Waste in the UK'. Final Report. Available at: http://www.wrap.org.uk/retail/case_studies_research/report_household.html It is estimated that 8.3 million tonnes per year of food and drink waste is generated by households in the UK. This is the equivalent to 330kg per year for each household in the UK, or just over 6kg per household per week. This figure is higher than previous estimates as it includes food waste disposed of in sewers, which is an estimated 1.8 million tonnes per annum.

p. 211 'as at least 40%' – Hall K. D., and Guo, J., et al. (2009), 'The Progressive increase of Food Waste in America and Its Environmental Impact', *PLoS ONE*, Vol. 4, Iss. 11, p. 7940.

p. 211 'for some crops' – Parfitt. J., Barthel, M., and Macnaughton, S. (2010), 'Food waste within food supply chains: quantification and potential for change to 2050', *Phil. Trans. R. Soc.*, B 2010 365, 3065–81.

p. 211 'Tristram Stuart argues' – Stuart, T (2009), *Waste: Uncovering the Global Food Scandal*, London: Penguin, p. 190.

p. 212 '7 million tonnes a year' – WRAP (2010). 'Waste arising in the supply of food and drink to households in the UK'. Available at: http://www.wrap.org.uk/retail/case_studies_research/report_waste.html

p. 212 'onto a plate' – UNEP (2009), The Environmental Food

Crisis. Available at: http://www.unep.org/publications/search/pub_details_s.asp?ID=4019

p. 212 'gets to us' – Parfitt, J., Barthel, M., and Macnaughton, S. (2010), 'Food waste within food supply chains: quantification and potential for change to 2050, *Phil. Trans. R. Soc.*, B 2010 365, 3065–81

p. 215 'gas requirements from biogas' – National Grid (2009), 'Half Britain's homes could be heated by renewable gas, says National Grid' [press release]. Available at: http://www.nationalgrid.com/corporate/About+Us/climate/press/020209.htm

p. 215 '4.3 tonnes of CO_2 equivalent' – Equivalent carbon dioxide (CO_2e) is a metric to assess the potential equivalent amount of CO_2 (and thereby an indication of the extent of CO_2 initiated radiative forcing and associated global warming) that a certain event or emission is responsible for, when the event or emission is not one that results in CO_2 entering the atmosphere. For example a methane emission.

p. 215 'emissions are prevented' – WRAP (2009), 'Household Food and Drink Waste in the UK', Final Report, pp. 92–94. Available at: http://www.wrap.org.uk/retail/case_studies_research/reports_household.html

p. 215 'just a tonne of CO_2 equivalent emissions' – Research on behalf of WRAP, demonstrated in personal communication from James, K., 11 August 2010.

p. 216 'shouted to warn passers-by' – Hart-Davis, A. (2006), *Just Another Day – the science and technology of our everyday lives*, London: Orion Books, p. 79.

p. 216 'exactly who was responsible' – Hart-Davis credits Joseph Bramah and then George Jennings: Rose George says Sir John Harrington (godson to Queen Elizabeth I) and later Alexander Cumming, Joseph Bramah and Thomas Crapper.

p. 216 'Dutch word meaning rubbish' – Hart-Davis, A. (2006), *Just Another Day – the science and technology of our everyday lives*, London: Orion – Books, p. 82.

p. 217 '5 million tonnes of CO_2 emissions' – 'The water industry is the fourth most energy-intensive sector in the UK': Parliamentary Office of Science and Technology (2007), 'Energy and Sewage: Postnote', briefing. Available at: http://www.parliament.uk/documents/post/postpn282.pdf And 'Without intervention, increased wastewater treatment under the Water Framework Directive (WFD) is likely to increase carbon dioxide (CO_2) emissions by over 110,000 tonnes per year from operational energy use and emissions associated with the additional processes required. This is a small increase with respect to the water industry's carbon footprint of five million tonnes (2007/2008), but the increase more than doubles the operational and capital emissions of individual works that require additional processes.' Environment Agency (2009), 'Transforming wastewater treatment to reduce carbon emissions', Resource Efficiency Programme. Available at: www.environment-agency.gov.uk/research/library/publications/114393.aspx

p. 217 'historian David Edgerton explains' – Edgerton, D. (2008), *The Shock of the Old – Technology and Global History since 1900*, London, Profile Books.

p. 218 'use of manufactured fertilisers' – Lloyd, J. in Hislop, Hannah, ed. (2007), *The Nutrient Cycle: Closing the Loop*, London: Green Alliance, diagram, p. 12. Available at: http://www.green-alliance.org.uk/uploadedFiles/Publications/reports/TheNutrientCycle.pdf

p. 218 'quality controls are in place' – DEFRA (2009), 'Attitudes to use of organic resources on land', research carried out by the Open University. Project WR0510.

p. 219 'simply isn't the money to build it' – Ecosan (2005), 'What is Ecological sanitation or Ecosan?' [website]. Available at: http://www.ecosan.nl/page/447

p. 219 'new toilets must be of this kind' – Tanum Municipality (2008), 'Urine Separation' [website]. Available at: http://www.tanum.se/vanstermenykommun/miljo/toaletterochavlopp/urineseparation.4.8fc7a71 04a93e5f2e8000595.html

p. 219 'their own food requirement' – Esrey, S., et al. (2001), 'Closing the Loop: Ecological Sanitation for food security', UNDP & SIDA, Mexico. Available at: http://www.ecosanres. org/pdf_files/closing-the-loop.pdf

p. 219 'go on the fruit trees' – CAT (2010), 'Compost Toilet', Centre for Alternative Technology [website]. Available at: http://www.cat.org.uk/vt/vt_content.tmpl?sku=VT_03/24

p. 220 'waterless (or nearly of water less) washing machines' – XEROS (2010) [product website]. Available at: http://www. xerosltd.com/.

p. 248 'Procter & Gamble' – Rathje, W., and Cullen, M. (2001), *Rubbish! The Archaeology of Garbage*, USA: University of Arizona Press, p. 154.

p. 221 'fifty-seven times its weight in water' – Richer Investment Diaper Consulting Services (2007), 'Frequently asked questions about disposable diapers'. Available at: http://www.disposablediaper.net/faq.asp?1

p. 222 'uptake is growing in the UK' – Smith, L. (2008), 'Green mums dump the disposable nappy to ease waste', *The Times*, 28 April.

p. 222 'around 3% of the household bin' – J.P. Parfitt and E. Bridgwater (2010) Municipal waste composition – what is still in the residual bin and what can we get out? Proceedings Waste 2010: Waste and Resource Management – Putting Strategy into Practice, Stratford-upon-Avon, 28–29 September 2010.

p. 222 'As much as 38%' – Burnley, S. (2007), 'A review of municipal solid waste composition in the United Kingdom', *Waste Management*, Iss. 27, pp. 1274–85.

p. 222 'nappies in household waste' – Jones, C. (2006), 'Rather than let pet dung go to waste, experts explore its energy potential', *San Francisco Chronicle*, 21 February.

p. 222 'does fully degrade' – Iwahashi, M. (2003), 'Mechanism for degradation of poly(sodium acrylate) by bacterial consortium', *Journal of Bioscience and Bioengineering*, Vol. 95, Iss. 5, p. 483.

p. 223 'has similar retentive properties' – Navarsage-Heald,

D. (2005), 'Thermal polyaspartate as a biodegradeable alternative to polyacrylate and other currently used water soluble polymers', University of Oregon.

p. 223 'in Cornwall' – see http://www.greenandpleasant.org.uk/

Chapter 16

p. 225 'Eating the Sun' – Morton, O. (2007) Eating the Sun – How Plants Power the Planet, Fourth Estate, London

p. 228 'who can rehearse them' – Mackay, F. (2008), Sustainable Energy – without the hot air, Cambridge: UIT. Also Collins, J. (2004), A Microgeneration Manifesto, Green Alliance. Available at: http://www.greenalliance. org.uk/grea_p.aspx?id=350 See also Cary, R. (2010), 'Future Proof: an electricity network for the 21st Century', Green Alliance. Available at: http://www.greenalliance.org.uk/ grea_p.aspx?id=4762

p. 230 'rather than a replacement' – The Tories think Britain could – with a fair wind – get 15% of its energy from decentralised sources by 2020, if financial incentives introduced by the government are applied more widely – Conservative Party (2010), 'An invitation to join the Government of Great Britain,' [manifesto] p. 191. Available at: http://www.conservatives.com/Policy/Manifesto.aspx

p. 231 'firm standards should be set' – Hislop, H. (2010), 'Tune in, Turn on, Cop out?', Business Green Comment. Available at: http://www.businessgreen.com/business-green/comment/ 2259248/tune-turn.cop

p. 231 'for an "energy-accredited TV' – DEFRA (2009), 'Leading UK retailers pledge to stop selling energy wasting TVs' [press release]. Available at: http://www.carbonchallenge.typepad.com/ carbon_challenge/2010/03/leading-UK-retailers-pledge-to-stop-selling-energy-wasting-TVs-.html

p. 231 'that boomed' – http://consumers.ofcom.org.uk/

2010/06/high-definition-brings-fans-closer-to-world-cup-action/

p. 232 '24 million tonnes of CO_2' – DEFRA (2009), 'Saving Energy Through Better Products and Appliances' [consultation] December. Available at: http://www.Defra.gov.uk/corporate/consult/ energy-using-products/index.htm

p. 233 'their manufacturing systems' – WRAP (2009), 'Meeting the UK climate change challenge: The contribution of resource efficiency'. Available at: http://www.wrap.org.uk/ wrap_corporate/annual_conference/resource_efficiency.html Also Peters, G. P., and Hertwich, E. G. (2008), 'CO_2 embodied in international trade with implications for global climate policy', *Environmental Science and Technology*, Vol. 42, iss. 5, pp. 1401–7.

p. 233 'Baghdad Battery' – Von Handorf, D. (2002), 'The Baghdad Battery: Myth or reality?', *Plating and Surface Finishing*, Vol. 89, No. 5, pp. 84–7.

p. 234 'converted to electrical energy' – Galvani, L. (1792), *Aloysii Galvani* . . . De Viribus Electricitatis in Motu Musculari Commentarius. Cum Joannis Aldini Dissertatione Et Notis. Accesserunt Epistolæ Ad Animalis Electricitatis Theoriam Pertinentes. Mutinæ: Apud Societatem Typographicam.

p. 234 'size of a coin powering cars' – Cadex Electronics Inc. (2010), *Batteries in a portable world – A handbook on rechargeable batteries for non-engineers*. Available at: http://www.buchmann.ca/

p. 234 '3% are presently recycled' – WRAP (2010), 'Battery facts & figures' [website]. Available at: http://www.wrap.org.uk/ local_authorities/research_guidance/collections_recycling/ batteries/battery _recycling_information/battery_ facts_html

p. 234 '45% by 2016' – European Commission (2009), Waste Batteries Directive. And WRAP (2009), 'Battery facts & figures'. Available at: http://www.wrap.org.uk/local_authori

ties/research_quidance/collections_recycling/batteries/battery
_recycling_information/battery_facts_html

p. 235 'catalysts from less expensive metals' – Atanassov, P., and
Olson, T. S. (2005), 'Non-Platinum Electrocatalysts for fuel
cells' [conference proceedings]. Available at: http://aiche.
confex.com/aiche/2005/techprogram/P24351.htm

p. 236 'been in development since 2008' – Harris, M. (2010),
'Toshiba's mobile phone fuel cell to appear at last?',
TechRadarUK [website]. Available at: http://www.techradar.
com/news/portable-devices/toshibas-mobile-phone-fuel-
cell-to-appear-at-last–623390

p. 236 'small phone chargers' – Maisto, M. (2010), 'Nokia
patent targets mobile device kinetic energy charging',
eWeek Europe [website]. Available at: http://www.eweek
europe.co.uk/news/news-mobile-wireless/nokia-patent-
targets-mobile-device-kinetic-energy-charging-5723

Chapter 17

p. 240 'for appliances and TVs 33%' – Cooper, T. (2004),
'Inadequate life? Evidence of consumer attitudes to
product obsolescence', *Journal of Consumer Policy*, Iss. 27, pp.
421–49, quoted in WRAP (2009), 'Meeting the UK climate
change challenge: The contribution of resource efficiency'.
Available at: http://www.wrap.org.uk/wrap_corporate/
annual_conference/resource_efficiency.html

p. 240 'less than 1% recycled' – SERI & FOEA (2009),
'Overconsumption? Our use of the World's Natural
Resources', September. Available at:
http://www.foeeurope.org/publications/2009/
Overconsumption_Sep09.pdf

p. 240 'clothing and other textiles' – DEFRA (2009),
'Maximising Reuse and Recycling of UK Clothing and
Textiles' [report summary]. Available at: http://randd.Defra.
gov.uk/Document.aspx?Document=EV0421_8745_FRP.pdf

p. 240 'remaking and reselling them' – Palmer, A., and Clark,

H. (2005), *Old Clothes, New Looks – Second Hand Fashion*, Oxford: Berg, p. 13.

p. 241 'a range of exotic textiles' – Palmer, A., and Clark, H. (2005), *Old Clothes, New Looks – Second Hand Fashion*, Oxford: Berg, p. 59.

p. 241 'traders from all over the world' – DEFRA (2009), 'Maximising Reuse and Recycling of UK Clothing and Textiles' [report summary]. Available at: http://randd. Defra.gov.uk/Document.aspx?Document=EV0421_8746_ OTH.pdf

p. 241 '65% to 40% over the last decade' – Textile Recycling Association (2005), 'The State of the UK Textile Reclamation Industry: a national diagnosis of the state of the industry' [website]. Available at: http://www.textile-recycling.org.uk/ index.html.

p. 242 'still capable of being restored' – Antiques are Green (2010), 'About Antiques are Green' [website]. Available at: http://www.antiquesaregreen.org/metadot/index.pl? id=2182

p. 243 'or mended a clock' – Crawford, M. B. (2009), *Shop Class as Soulcraft: An Inquiry into the value of work*, London: Penguin.

p. 243 'gratification and disposability' – Sennett, R. (2006), *The Culture of the New Capitalism*, USA: Yale University Press.

p. 243 'over 1 million tonnes by 2015' – GLA (2010), '£8m to create UK's first city-wide reuse and repair service' [press release], 12 July. Available at: http://www.legacy.london. gov.uk/assembly/assemmtgs/2010/mqtsep15/item03.pdf

p. 244 'from London annually' – http://www.environment- agency.gov.uk/research/library/publications/41039.aspx

p. 245 'that number of people' – Papanek, V. (1995), *The Green Imperative – Ecology and Ethics in Design and Architecture*, London: Thames and Hudson.

Chapter 18

p. 246 'overseeing its regulation' – see http://www.defra. gov.uk/acre/

p. 247 'glow in the dark' – see http://www.jellyfishfacts. net/glow-in-the-dark-jellyfish.html

p. 248 'seminal ecological trials' – see http://www.nerc.ac.uk/ press/releases/2003/21a-gmo.asp

p. 250 'car polish and tennis balls' – http://www.ranotech project.org/inventories/consumer

p. 250 'ultra-thin film of gel' – European Commission (2009), 'Nanomaterial Roadmap to 2015 – SWOT analysis concerning the use of nanomaterials in the Energy Sector', European Commission Report. Available at: http://www. nanoroad.net/download/swot_e.pdf

p. 250 'means of generating colour' – Hogan, J. (2004), 'Smart Surfaces show their Colours', New Scientist, 28 February, Iss. 2436.

p. 250 'devoted to space tethers' – The Space Elevator Reference (2009), 'The Vision' [website]. Available at: http://www.spaceelevator.com/

p. 251 'woven into fibre' – Dunn, J. (2009), 'Space Elevator ... And the next floor is outer space', The Times, 18 January. Available at: http://www.timesonline.co.uk/tol/driving/ features/article5529668.ece.

p. 251 'than aluminium' – Kushnir, D. and Sanden, B.A. (2008), 'Energy requirements of Carbon Nanoparticle production.' Journal of Industrial Ecology, Vol. 12, Iss. 3, pp. 360–75.

p. 251 'water intensive' – Plepys, A. (2004), 'The environmental impacts of electronics: Going beyond the walls of semicon-ductor fabs'. In 2004 IEEE International Symposium on Electronics and the Environment, IEEE, Piscataway, NJ.

p. 251 'persist in the environment' – Royal Commission on Environmental Pollution (2008), 'Novel Materials in the Environment: The case of nanotechnology', RCEP reports. Available at: http://www.rcep.org.uk/reports/ 27-novel%20materials/27-novelmaterials.htm

Chapter 19

p. 257 'only takes place in Sweden' – Tetra Pak (2007), 'Comparative Life Cycle Assessment of Tetra Pak Packaging Synthesis', BIO Intelligence Service. Available at: http://www.tetrapak.com/se/Documents/TetroPak_ACV_Synthese_UK_vf_2007%5B1%5D.pdf

p. 258 'for packaging materials' – BRC (2009), 'Retailers Launch New Recycling Label', British Retail Consortium News [website]. Available at: http://www.brc.org.uk/details04.asp?id=1531

p. 260 'sculpture at the Eden Project' – BBC (2006), 'Weee'll Be Back'. Available at: http://www.bbc.co.uk/cornwall/content/articles/2006/11/29/weeeman_feature.shtml [accessed 28 July 2010].

p. 260 'have to raise their game' – Environment Analyst (2008), 'Tougher WEEE collection target poses challenge to UK' [website]. Available at: http://environment-analyst.com/1001

p. 264 'order of things' – Rees, J. (1990), 'Natural Resources – Allocation, Economics and Policy', Routledge, p 434.

Chapter 20

p. 269 'than owning it' – http://www.treehugger.com/files/2008/11/good-product-service-systems.php

p. 272 'Edgerton has made clear' – Edgerton, D. (2008), *The Shock of the Old – Technology and Global History Since 1900*, London, Profile Books.

p. 274 'Gordon Brown when Chancellor' – for CEMEP report see http://www.defra.gov.uk/environment/business/innovation/commission/documents/cemep-report.pdf

p. 278 'the CoolProducts campaign' – CoolProducts (2010), 'CoolProducts: for a cool planet' [home page]. Available at: http://www.coolproducts.eu/

p. 280 'although not the reasons for this' – OPRL (2010),

'Retailers and Brand Owners – Current Members'. Available at: http://www.onpackrecyclinglabel.org.uk/default.asp?section_id=2&content_id=6

p. 281 'taking shape in France' – see http://www.agrion.org/world_debate/agrion-enEco_labeling_countdown_to_compliance.htm

p. 283 'buying recycled paper' – 'Central civil government spends some £13 billion a year on goods and services, including over 21,000 tonnes of copier paper. A government wide commitment to buy recycled paper will, for example, save some 350,000 trees, enough energy to heat 10,000 average homes, and 300 million gallons of water every year' – DEFRA, OGC & Buying Solutions (2010), 'Government buys into sustainable procurement' [press release]. Available at: http://www.ogc.gov.uk/7023_4188.asp

p. 283 'FSC-certified timber' – 'Departments should note that, in line with the commitment in the UK's Sunstainable Action Plan, the Government's policy on timber procurement was updated on 1 April 2009. Central departments, their executive agencies and non depart-mental public bodies must now only procure timber and wood derived products originating from either legal and sustainable or FLEGT licensed sources. Local government is not required to follow this policy, but it is encouraged to comply' – OGC (2009), 'Timber Procurement', OGC Policy Principles. Available at: http://www.ogc.gov.uk/documents/Timber Policy%281%29.pdf

p. 283 'abandoning bottled water' – DEFRA (2009), 'Selling to Defra: Guide for Suppliers'. Available at: http://www.Defra.gov.uk/corporate/about/how/procurement/documents/suppliers-guide.pdf

p. 283 'being managed at the time' – http://www.ogc.gov.uk/gps_digest_procuring_a_zero_waste_mattress_and_pillow_solution_for_hm_prison_service.asp

p. 285 'good and bad products' – Hill J. et al (2008) *Good Product, Bad Product – making the case for product levies*

London, Green Alliance. Available at: http://www.green-alliance.org.uk/grea_p.aspx?id=2730

p. 285 'VAT on the EU agenda' – European Commission (2005), 'The use of differential VAT rates to promote changes in consumption and innovation', summary, 25 June. Available at: http://ec.europa.eu/environment/enveco/taxation/pdf/vat _summary.pdf 'A number of EU Member States already have some experience with using reduced VAT rates to promote environmentally preferred products. For example, from 1993 until 2004 the Czech Republic applied reduced VAT rates to a number of products including renewable energy equipment, biofuels and recycled paper. In Portugal, equipment necessary for the production and use of renewable energy resources is taxed at a 12% VAT rate (instead of 21%).'

p. 285 'A carbon tax' – GFC (2009), 'Final Report: The Case for Green Fiscal Reform', Green Fiscal Commission. Available at: http://www.greenfiscalcommision.org.uk/images/uploads/GFC_FinalReport.pdf

p. 286 'total aggregate used'. Budget (2009), HC 407 April 2009, para. 7.67. Available at: http://www.hmrc.gov.uk/budget 2009/index.htm.

p. 287 'these ideas' – for the report from the first two years of the Designing out Waste consortium, see http://www.green-alliance.org.uk/grea_p.aspx?id=5016

p. 292 'not selling patio heaters' – Brignall, M. (2008), 'B&Q to end sale of patio heaters', Guardian, 22 January. Available at: http://www.guardian.co.uk/environment/2008/jan/22/carbonemissions.climatechange [accessed 21 June 2010].

p. 293 'ecological macroeconomics' – Jackson, Tim (2009) Prosperity without Growth: Economics for a finite plant, Earthscan. The key text of ecological macroeconomics is: Daly, H.E. ed. (1973), Towards a Steady-state Economy, W.H. Freeman & Co Ltd.

p. 294 'which makes life worthwhile' – Sen. Robert F. Kennedy (1968), speech at the University of Kansas, 18 March, JFK Presidential Library & Museum. Available at: http://www. jfklibrary.org/Historical+Resources/Archives/Reference+Desk/

Speeches/RFK/RFKSpeech68Mar18UKansas.htm. The full quote: 'We will find neither national purpose nor personal satisfaction in a mere continuation of economic progress, in an endless amassing of worldly goods. We cannot measure national spirit by the Dow Jones Average, nor national achievement by the Gross National Product. For the Gross National Product includes air pollution, and ambulances to clear our highways from carnage. It counts special locks for our doors and jails for the people who break them. The Gross National Product includes the destruction of the redwoods and the death of Lake Superior. It grows with the production of napalm and missles and nuclear warheads ... it includes ... the broadcasting of television programs which glorify violence to sell goods to our children. And if the Gross National Product includes all this, there is much that it does not comprehend. It does not allow for the health of our families, the quality of their education, or the joy of their play. It is indifferent to the decency of our factories and the safety of our streets alike. It does not include the beauty of our poetry, or the strength of our marriages, the intelligence of our public debate or the integrity of our public officials ... the Gross National Product measures neither our wit nor our courage, neither our wisdom nor our learning, neither our compassion nor our devotion to our country. It measures everything, in short, except that which makes life worthwhile, and it can tell us everything about America – except whether we are proud to be Americans.'

p. 294 'philosopher Kate Soper' – Soper, K., and Thomas, L. (2006), '"Alternative Hedonism" and the Critique of "Consumerism"', Working Paper, London Met, referenced in Jackson, T. (2009), *Prosperity without Growth: Economics for a finite planet*, Earthscan.

p. 296 'detailed ones afterwards' – Bowker, R. (2007), 'Children's perceptions and learning about tropical rainforests: an analysis of their drawings'. *Environmental Education Research*, Vol. 13, No. 1, February, pp. 75–96.

Index

ACCPE (Advisory Committee on Consumer Products and the Environment), 155-6, 157
acid drainage, 56-7
AD *see* anaerobic digestion
Adams, Douglas: *Last Chance to See* (co-written with Mark Carwardine), 92-3
additives, 12, 25, 63, 202
Advertising Standards Authority (ASA), 155
Advisory Committee on Consumer Products and the Environment *see* ACCPE
aerogel, 180
aerosol loading, 119
'affluenza', 131
Africa, 56, 139, 271
   West, 97
aggregates, 49-50, 54, 194, 286
Aggregates Levy, 285-6
agriculture, 38, 77, 90, 91, 93, 95, 96, 97, 122, 208-9
   subsidies, 211
air-source heat pumps, 229
algal blooms, 143
aluminium, 48, 55, 58-9, 60, 169, 180, 250, 256, 258, 260
Amazon, 197
amino acids, 35
anaerobic digestion (AD), 16-17, 214, 215, 219, 222-3
Antarctica, 106
antiques, 242

Antiques are Green campaign, 242
Apollo 8, 41
Apollo 11, 41
Aral Sea, 98
Arizona, 97
arsenic, 57, 104-5
ASA (Advertising Standards Authority), 155
Ashby, Michael, 48, 80
Asia, 39, 146, 271
   South-East, 59
atoms, 36
Australia, 59, 94, 97
Austria, 27, 196
Azerbaijan, 40
Aztecs, 38

Babylonians, 198
Baekeland, Leo, 70
Baghdad battery, 233
Bakelite, 70
Baltics, the, 150
Ban, Shigure, 149-50
bananas, 153-4, 248-9
B&Q, 156, 292
banking crisis/credit crunch (2008), 113-14, 117, 266
Basic Law for Establishing the Recycling-Based Society (Japan), 172
Basle, 64
batteries, 5, 6, 11, 233-5, 237, 239
bauxite ore, 59
Bedfordshire, 19

beeswax, 9
behavioural economics, 133, 182-3
Behrens, Arno, 118
Beijing, 91, 102
Benn, Hilary, 87, 108-9
Benyus, Janine, 203
Berkeley Pit, 104
Berners-Lee, Mike, 198
    How Bad Are Bananas?, 153
Bessemer process, 40
Biffa, 29-30, 289
Big Bang, 34
big kit, 228, 229
biodegradable materials, 16-17,
    20, 25, 203, 205 see also
    nutrients cycled
biodiversity loss see extinctions/
    biodiversity loss
biofuels, 208-9
biogas, 214, 215, 219, 270
biomass energy (plant-based
    energy), 225-6
biomimicry, 202-3
bioplastics, 204-9
biorefineries, 209, 246, 249
Bissel, George, 40
Black, Maggie: No-Nonsense Guide
    to Water, 96
bleach, 272
'blood minerals', 112
Bog Standard website, 146
Bolton, Giles: Aid and Other Dirty
    Business, 139
boilers, 229
Bonington, Sir Chris, 17
books, 197, 198, 199
bottle banks, 53
Boulding, Kenneth, 23, 25
boundaries, planetary, 119-20
BP, 111-12
Brand, Henning, 35-6
Braungart, Michael: Cradle to
    Cradle (co-written with
    William McDonough), 201-2,
    210
Brazil, 57

bricks, 49
Brighton: materials recovery
    facility, 255-9
Britain see UK/Britain
British Agrochemicals
    Association, 277
British Antarctic Survey, 106
British Empire, 129
bronze, 38
Bronze Age, 37, 76
Brown, Gordon, 274, 285
Brown, Lester, 115
Bryson, Bill: A Short History of
    Nearly Everything, 92
building materials see construc-
    tion materials
businesses, 286-8 see also
    producer responsibility
Butte, 104-5

Cadbury Schweppes, 24
cadmium, 55, 57, 104
calcite, 36
calcium, 47, 48
California, 77, 192, 275
Cambodia, 62
Cameron, Dr J.A., 147
Canada, 59, 61, 65, 66, 111, 273
cans, 60, 169, 256, 258
capitalism, 128, 132
carbohydrates, 204
carbon, 33, 36, 91, 99, 150, 208, 210
    emissions, 119, 215, 217, 228
carbon capture and storage
    systems, 238
carbon dioxide, 91, 100, 119, 150,
    208, 215, 226, 238
carbon fibres, 80, 233
carbon footprints, 152, 153, 154,
    279
carbon nanotubes, 250-1
carbon tax, 285
Carbon Trust, 152
care with new promises, 185,
    246-52
Caribbean, 217

carpet, 79
cars, 260, 262-3
  electric, 182
Carwardine, Mark: *Last Chance to See* (co-written with Douglas Adams), 92-3
cashmere, 75
cellulose, 204-5
cement, 48, 49, 121
CEMEP (Commission on Environmental Markets and Economic Performance), 274
Centre for Alternative Technology, 219
'ceramics', 48-9, 51 *see also* names of materials
CFCs, 106, 261
charity shops, 137, 138, 241
chemicals, 64-5, 67, 102-3, 106, 121, 261
chemistry, 35-6, 37, 70, 71
China
  American demand for cheap goods from, 170
  blast furnace reaches Europe from, 40
  cement use, 194
  industrial expansion, 27-8
  and manufacturing technology, 273
  market for recycled materials, 28, 68, 73, 173
  and metals, 59, 113
  and paper, 64, 65, 66, 68
  phosphate reserves, 145
  pollution, 101, 102
  and sanitation, 146, 219-20
  skill with clay, 50-1
  brief references, 11, 271
china, 50-1
china clay, 51
chipboard, 63
chlorine bleach, 272
choice-editing, 156, 292
chromium, 47
Clark Fork River, 104

clay, 50-1
Clean Air Act (1956), 102
climate change, 91, 95, 99-101, 119, 121 *see also* global warming
Climate Change Act, 123
'climax' ecosystem, 61
closed loop, 67, 74, 203, 209, 215, 219, 258-9
Closed Loop plant, Dagenham, 74
'Closing the Loop' project (Green Alliance), 206
clothing, 10, 76, 78-9, 240, 241-2 *see also* textiles
Club of Rome, 109
coal, 40, 47-8, 56, 117-18, 227, 238
coal technology, clean, 238
co-cropping, 209
coffee, 168-9, 180-1
Colorado River, 97
coltan, 112
Commission on Environmental Markets and Economic Performance (CEMEP), 274
Committee on Climate Change, 183
compliance schemes, 262, 263
composites, 79-81, 196, 279
compost, peat in, 151-2
composting, 9, 11, 12, 16, 174, 206-8, 280
composting toilets, 219, 220
computers, 41, 97
concrete, 39, 48-50, 96-7, 194-5, 196
Constanza, Robert, 120
construction materials, 48-50, 82, 96-7, 194-7
consumption, 22, 29-30, 78, 113-14, 123, 127-38, 140, 183, 292-4
  of water, 96-8
cool products, 228
Coolproducts campaign, 278
Cooper, Tim, 239-40
Copenhagen Amendments (1992), 106

Copenhagen climate talks (2009), 100, 286
copper, 56, 57, 104, 111, 188, 260, 262
co-products, 190, 209
Cornwall, 194 *see also* Eden Project
cotton, 10, 75, 76, 77, 78, 97, 98, 99, 121, 149, 190-1
cradle-to-cradle, 201-2, 288-9
Crapper, Thomas, 216
Crawford, Mathew, 243
credit crunch/banking crisis (2008), 113-14, 117, 266
Crick, F.H.C., 35
Crimplene, 75-6
crisps, 152, 279
    packets, 152, 258
Croft House Museum, Shetland, 162
crops
    energy from, 225-6
    genetic modification of, 247-8
    see also agriculture
cross-laminated wood, 195-6
Cumbria, 95

Dagenham: Closed Loop plant, 74
Dalton, John, 36
Daly, Herman, 114, 115
Dark Ages, 39
Darling, Alistair, 274
Darwinism, 35
de Boton, Alain, 130, 131, 132, 133-4
    Status Anxiety, 130
DECC, 183
deconstructing your home, 42-6
deforestation, 90-1, 116-17
DEFRA (Department for Environment, Food and Rural Affairs), 151, 154-5, 156, 283
de Havilland, 193-4
de-inking waste, 67
demand, 28, 117, 118, 224, 225

Democratic Republic of Congo, 112
Denmark, 21, 52
Department for Environment, Food and Rural Affairs (DEFRA), 151, 154-5, 156, 283
Department of Health, 99-100
designing the future, 177-252
'Designing out Waste' project (Green Alliance), 206
detergent, 142-6
Diamond, Jared, 139-40, 274
    Collapse, 115-17
diamonds, 112, 188-9
diesel, 40
digestate, 214
digesters *see* anaerobic digestion
digital storage, 199
dish-washing products, 145
DNA, 35
Dow, Chris, 74
Dragons' Den, 4-5, 7
Drake, Edwin, 40

Easter Island, 116
eBay, 241, 242
e-books, 198
Eco-Build exhibition, London (2010), 195
eco-labels, 154
ecological economics, 114-15, 293
ecology, 36
economic growth, 22, 83, 108-9, 110, 113-14, 131, 160, 293
economics, 25, 28, 111-12, 114-15, 118, 128, 131, 133, 211-12, 264, 270-1, 293-4; *see also* economic growth; linear economy
ecosan (ecological sanitation), 219, 220-1
eddy current separator, 256, 261
Eden Project, 6, 34, 57, 64, 188, 203, 230, 260, 294-7
Edgerton, David, 217, 272, 273
Edward I, King, 102
efficiency, 117-18, 122, 200-1

energy, 118, 225, 231, 232, 233, 278
Egyptians, 38, 64
electric cars, 182
electricity, 227-8, 230, 234, 235
electroluminescence, 6
electronic book readers, 198
electronic media, 198
elements/elemental forces, 34-7
emails, 198
embodied energy, 232-3
End of Life Vehicles Directive, 263
energy
  and climate change, 98-101
  efficiency, 118, 225, 231, 232, 233, 278
  and entropy, 25, 224
  in imagined future, 8, 269-70
  and incineration, 21, 72, 164
  and Jevons paradox/rebound effect, 118, 125
  and limits, 122
  moving towards renewable, 185, 215, 224-38
  and production process, 59, 77, 98-9, 121, 152, 194, 198, 205, 278-9
  and product standards, 277-8, 278-9
  and recycling, 26, 54, 67, 72, 239
  and sewage treatment, 217
  supply and demand, 224-5
  and taxation, 285
energy harvesting, 236
energy labels, 231
England, 8, 15, 19, 61, 91, 95, 99, 265 see also UK/Britain
entropy, 24-6, 74, 200, 224
Environment Agency, 6
environmental indicators, 281
environmental literacy, 294-7
environmental movement, 23-4, 26-7
environmental taxes, 284-5
erosion, soil, 94, 116-17

Europe, 27, 40, 62, 72, 101, 128, 143, 145, 147, 150, 205, 214, 219, 236, 247-8, 276 see also European legislation; European Union; names of countries
European legislation, 101, 214, 215, 260
European Union (EU), 20, 53, 144, 214, 215, 217, 231, 232, 259, 276, 278, 285, 288
  Batteries Directive, 234
  Waste Electronic and Electrical Equipment (WEEE) Directive, 6, 263
  see also European legislation
eutrophication, 143, 144
Everest base camp, 17
evolution, 35
exploration, age of, 39
extinctions/biodiversity loss, 89, 91-3, 119, 121

fabrics see textiles
Fairtrade, 189
Farman, Joe, 106
feed-in tariffs, 230
fertilisers, 48, 143, 217-18
fertility, male, 103
fibre-glass, 195
Fim, Ken: My Journey with a Remarkable Tree, 62
Finland, 65, 198
fire, 37
fish, 212
flax, 76
Florence, 240
Florida, 88
flush toilets, 216, 219
fly-tipping, 17
food
  GM, 248-9
  and land use, 90, 122, 209
  prices, 209, 213
  waste, 16-17, 22, 72, 167, 210-15

Food Ethics Council, 193
Forest Disclosure Project, 62
forests, 39, 61, 62, 65-6, 90-1, 116-
    17, 121, 148, 196
Forest Stewardship Council *see*
    FSC
Forrester, Jay, 120
forward commitment procure-
    ment, 283
fossil fuels, 47-8, 99, 111-12, 225
    see also coal; oil
France, 281
'freecycle', 240
fridges, 261
Friends of the Earth, 23-4, 30,
    240, 277
Fry, Art, 181
Fry, Stephen, 92
FSC (Forest Stewardship Council),
    63, 147-8, 149, 154, 187, 282
fuel cells, 235-6
future, the
    designing, 177-252
    vision of, 81-2, 266-72

Gallagher Review, 209
Galvani, Luigi, 234
Ganges, River, 102
Gardiner, Brian, 106
gas, 19, 20, 227
genetic modification (GM), 246-9
George, Rose, 146-7, 216, 219
    The Big Necessity, 216
geothermal energy, 226, 228, 230
Germany, 27, 214, 217-18, 219,
    228, 230
Ghent Virtual Bakelite Museum,
    70
Girling, Richard, 15, 130, 133, 134
    Greed, 130
    Rubbish, 130
glass, 7, 9, 10, 22, 25, 27, 51-4,
    121, 257
glasses, light-up disposable, 5, 6-7
glass fibres, 80
globalisation, 212

global warming, 26, 95, 99-101,
    121, 137, 273
GM (genetic modification), 246-9
gold, 55, 56, 57, 112-13, 180, 189,
    262
Goleman, Daniel, 158-9, 289
'GoodGuide' (Internet service),
    159
Google, 198
government, 267-8, 269, 270, 275-6,
    282-4, 284-6, 287, 288, 289
Greeks, ancient, 15, 38
Green Alliance, 5-6, 206, 207, 228,
    229, 277, 278, 284, 285, 287
green claims code, 155
green consumers, 138, 139-61,
    290-1
green innovation forcing, 274-5
Greenland, 113
green procurement, 267-8
Gross National Product, 294
ground-source heat pumps, 229
guano, 217
Gulf of Mexico, 112

Haggith, Mandy: *Paper Trails*, 66
Hailes, Julia, 140
    Green Consumer Guide, 140
Haiti, 91
Halley Bay, 106
hardwoods, 62
Hart-Davis, Adam, 233
Hayam town, 175
HDPE (high density polyeth-
    ylene), 72-3, 257, 258
health, impact of climate change
    on, 99-100
heap-leaching, 57
heat-storage devices, 229
helium, 34
hemcrete, 196
hemp, 76, 196
Hino City, 174-5
Hislop, Hannah, 207
Holland *see* Netherlands/Holland
hot spots, 192, 279

human waste *see* sewage
hydroelectric power, 228
hydrogen, 34, 235-6
hydrogenation, 217-18

ICMM Sustainable Development
    Principles, 57
incentives, 291
incineration, 20-1, 164, 171
India, 101, 102, 191, 271
Indonesia, 66
Industrial Revolution, 39-40, 129
industrial waste, 18, 23, 27, 28,
    82
insecticides, 77
Ireland, 150
iron, 39, 47, 48, 55, 104
Iron Age, 37, 38
Israel, 24, 97
Italy, 57

Jackson, Tim, 135
    Prosperity without Growth,
        293-4
James, Oliver, 130, 131, 132, 133-4
    Affluenza, 130
Japan, 104, 128, 149, 171-5, 189,
    216, 240, 273
Japan Containers and Package
    Recyling Association, 172-3
jatropha, 192
Jevons paradox, 117-18, 193, 225,
    270, 273-4
jewellery, 188-9
Johnson, Boris, 243-4
Johnson & Johnson, 221
Joyce, James: *Ulysses*, 146

Kamikatsu, 171-5
Kasamatsu, Kazuichi, 174
Kazakhstan, 98
Kennecott mine, 188
Kennedy, Robert, 294
Kenya, 139
kerosene, 40
Keruing, 62

Keynes, Maynard, 131
King's College, London, 179-80
Knight, Alan, 155-6, 292

labelling, 147, 151, 152-3, 154-5,
    156, 183, 193, 206, 231, 279,
    280-1
land, 88-9, 90-1, 119, 122, 208-9
landfill, 7, 18-20, 21, 23, 50, 53, 79,
    82, 189-90, 214-15, 222
landfill tax, 284-5
landslides, 91
Landy, Michael, 43-4, 45
Large Hadron Collider, 41
lash-ups, 181-2, 183, 184, 237,
    244, 272
Las Tablas de Daimiel national
    park, 151
Laughlin, Dr Zoe, 179-80
Lavoisier, Antoine and Marie, 36
Lawson, Neal, 130-1, 132-3, 133-4
    All Consuming, 130-1
LDPE (low density polyethylene),
    72, 73
leachate liquid, 18-19
lead, 39, 55, 57, 104
Leeds University, 113
Le Guin, Ursula: *The Dispossessed*,
    136-7
Leonard, Annie: *The Story of Stuff*,
    158
Lerwick, 164
Lesotho, 94
Liberia, 112
Life Laundry, 43
lightbulbs, 162, 230-1
lignin, 204-5
lime, 196
limits, 107, 108-23, 160, 292-3
'Limits to Growth' report (1972),
    24, 109-10, 114
linear economy, 23, 29, 51, 54, 58,
    63, 69, 74, 78, 82, 242-3
lithium, 234
Lloyd, Christopher: *What on Earth
    Happened*, 37

local councils/authorities, 13, 14, 15, 22-3, 29, 68, 73, 207, 259, 262, 269, 280

London, 43-4, 101-2, 220-1, 244
Eco-Build exhibition (2010), 195
Science Museum: '100 Years of Making Plastics' exhibition, 69-70

London Reuse Network, 244

Lunn, Pete, 130, 133, 135
Basic Instincts, 130, 133

McDonough, William: Cradle to Cradle (co-written with Michael Braungart), 201-2, 210

Machiavelli, 264

MacKay, David, 230, 232-3

Made-By, 191

Mad Men, 129

magazines, 66

magnesium, 47

Maikelm, Julian, 223

Maldives, 88

manganese, 104

marine litter, 163

markets, 4, 24, 26, 27-8, 73, 172, 173, 229, 260

Massachusetts Institute of Technology, 109

mass-balance, 29-30, 289

materialism, 132

materials recovery facility (MRF), 255-9

Mayans, 15, 38, 116-17

MDF, 63

Meadows, Dennis, 114

meat consumption, 211

mechanisation, 39-40

Mendeleev, Dmitri, 36

mercury, 55, 56, 104

metals, 11-12, 22, 33, 34-5, 37, 38-9, 47, 54-60, 99, 103-5, 110-11, 112-14, 180, 270
recycling, 27, 59-60, 81-2, 256, 258, 260-3

stewardship for, 187-90
see also names of metals

methane, 20, 211, 215, 222

Met Office, 95

Mexico, 116

microchip production, 251

micro-generation, 228, 229

Middle East, 24, 39

Miliband, David, 274

Miller, Daniel, 134-5
The Comfort of Things, 134-5
Stuff, 134-5

minerals, 34-5, 36-7, 45, 47-8, 52, 110-14, 121 see also metals; names of minerals

Minimata Bay, 104

mining, 55-8, 97, 104, 187-8

Ministry of Supply, 22

mobile phones, 11, 159-60, 236, 240

Modal, 77

Molotch, Harvey, 181-2, 184, 244, 272

Mongolia, 94

Montana, 104-5

Montreal Protocol, 106, 286

Moore, Charles, 74-5

Morton, Oliver, 225, 226, 227
Eating the Sun, 225

Mosquito aircraft, 193-4

mottanai, 171

MRF (materials recovery facility), 255-9

Murano, 52

Murray-Darling River system, 97

mussels, 203

Naish, John, 130, 131, 133-4
Enough, 130

nanotechnology, 249-52

nappies, disposable, 221-3

naptha, 71

NASA, 106, 180

National Academy of Sciences (US), 115

National Grid, 215

National Parks (US), 169-70
National Trust, 151
natural capital, 88-9
    running down, 114-18
'needs' and 'wants', 136-7
nef (new economics foundation),
    122
    'Growth isn't possible' (report
        2010), 109, 115
neodymium, 113
Neolithic stone homes, 49
Netherlands/Holland, 21, 88, 153,
    192, 214
nettles, 76
neurodevelopmental disorders,
    103
neurotoxins, 103, 104
new economics foundation see nef
New Scientist, 119-20, 199
newspapers, 68, 157
New Zealand, 92
NHS, 282
nickel 'super alloy', 180
niobium, 55
nitrate fertiliser, 218
nitrogen, 210
    cycle, 119
North America, 143, 145 see also
    Canada; US
North Sea, 112
Northwest Environment Watch:
    Stuff, 157-8
Nouvelle toilet tissue website, 146
nuclear power, 238
nutrients cycled, 185, 210-23
nylon, 77

oceans, 74-5
    acidification, 119
oil, 9, 24, 40, 47-8, 70-1, 77, 111-
    12, 227, 273
    crisis (1973), 24
optical sorter, 256-7
ores, 38, 55-6, 57
organic certification (textiles), 191
Orkney Islands, 49, 164

outsourcing, 122, 233
over-irrigation, 94
Oxfordshire, 50
oxygen, 36
ozone depletion, 105-7, 119

Pacific Islands, 217
Pacific Ocean, 75, 121
packaging, 11, 22, 42, 59, 71-2,
    152, 250, 258, 263, 280-1
Packard, Vance: The Waste Makers,
    129-30
Pampers, 221
Papanek, Victor, 244-5
paper, 22, 25, 27, 28, 45, 63-9, 82,
    97, 154, 197-9, 256, 258, 282 see
    also toilet paper
papyrus, 64
Pawlyn, Michael, 203
PE (polyethylene), 71, 72-3
Pearce, Fred, 97, 98
    Confessions of an Eco Sinner, 59,
        158
peat, 150-2
PEFC (Programme for the
    Endorsement of Forest
    Certification), 63, 147
penalties, 291-2
Pennsylvania, 40
periodic table, 36
persistent organic pollutants
    (POPs), 103
Peru, 217
pesticides, 277
PET (polyethylene terephthalate),
    71, 72, 74, 257, 258, 280
petrol, 40, 70-1
Petty, Kate: Earthly Treasures, 34
Phoenix, Arizona, 97
phospate, 48, 142-6, 210, 218
phosphate-stripping, 144-5, 218
phosphorus, 36
    cycle, 119
photosynthesis, 225
photovoltaic cells/panels, 183-4,
    226, 227, 229

physics, 41
piezoelectric, 236
planetary boundaries, 119-20
plant-based energy (biomass energy), 225-6
plastic, 6, 7, 10, 25, 28, 40-1, 45, 69-75, 80, 82, 97, 99, 121, 172-3, 180, 204, 257, 258, 279 see also bioplastics
Plastics Europe, 71
platinum, 55, 111, 235
pollution/pollutants, 26-7, 57, 75, 89, 101-5, 119, 121, 142-3, 218, 251
polyester, 77
polyethylene (PE), 71, 72-3
    high density (HDPE), 72-3, 257, 258
    low density (LDPE), 72, 73
polyethylene terephthalate (PET), 71, 72, 74, 257, 258, 280
polymers, 70, 71, 73, 74, 77, 78, 204 see also plastic
polypropylene (PP), 71, 73, 258
polystyrene, 258
polyvinyl chloride (PVC), 70, 71, 73, 74
POPs (persistent organic pollutants), 103
porcelain, 51
Post-it notes, 181
pottery, 50-1
pozzolana, 39
PP (polypropylene), 71, 73, 258
Premier, 112
pressure groups, 291
prices, 182-3, 275
    clothes, 241-2
    food, 209, 213
pricing out waste, 284-6
Prison Service, 283
Procter & Gamble, 221
procurement, 267-8, 282-4
producer responsibility, 262, 263, 288-90
product road maps, 156-7

product standards, 277-82
product stories, 155-6, 157, 159, 187
Programme for the Endorsement of Forest Certification (PEFC), 63, 147
prosperity, concept of, 294
protected stream, 188
PS, 73
PVC (polyvinyl chloride), 70, 71, 73, 74

quarries, 50

radical transparency, 158, 159, 289
RAF, 193
rainfall, 90-1, 95, 121
rare-earth metals, 55, 113, 189
Rathje, William: Rubbish!, 14-15, 221
raw material tax, 285-6
rebound effect see Jevons paradox
recycling, 11, 12, 13, 22, 23-4, 25-6, 26-9, 81-2, 165, 169, 172-3, 200-1, 255-65, 269, 278, 279-80
    batteries, 234
    composites, 80, 279-80
    designing for recovery, 185, 200-9
    glass, 53-4
    metals, 59-60, 81-2, 256, 258, 260-3
    paper, 67, 68, 82, 256, 258
    plastic, 72-4, 82, 257, 258, 279-80
    textiles, 78, 82
    wood, 63
Rees, Judith, 264
repair, 242-3
Responsible Jewellery Council (RJC), 188-9, 282
reuse, 21-2, 26, 52-3, 81, 165-6, 239-42, 243-4
Rio Tinto, 188
RJC (Responsible Jewellery Council), 188-9, 282

rocks, 49-50
Romans, 39, 49, 104, 142, 146
RSPB, 143
rubber, vulcanised, 79
Russia, 59, 61, 66, 271
Rwanda, 112, 139

saline intrusion, 98
salt contamination, 94
San Francisco, 222
sanitation, 215-21
Sarkozy, President, 285
Scandinavia, 59, 61
scandium, 55
Scotland, 49, 65, 265 see also
   UK/Britain
Scott Brothers, 146
second-hand commerce, 240, 241
Sennett, Richard, 243
sewage, 101, 102, 142-3, 144, 145,
   215-21
Shanklin, Jonathan, 106
sharing, 244-5
Shetland, 162-6
Shetland Amenity Trust, 163
Shetland Islands Council, 164-5
shikkui, 201
shoddy, 78
shopping, 129, 131, 137-8, 167
shower heads, 232
Sierra Club: Material World, 42-3
silicon, 41, 48
silicon chips, 192, 250
Silicon Valley, 105, 192
silk, 75, 76
silly putty, 180
silver, 55, 262
sludge, sewage, 217, 218
smart materials, 251-2
smelting, 59
sodium acrylate, 221, 222
softwood, 62
soil, 89, 93-5
solar energy, 8, 225, 226-7 see also
   solar panels; photovoltaic
   cells/panels

solar gain, 8
solar panels, 183-4, 229, 239
   tiny, 235
Soper, Kate, 294
South Africa, 59
South America, 56, 271
South-East Asia, 59
Soviet Union, 98
space, junk in, 17
space tethers, 250-1
Spain, 150, 151, 228, 230
species extinctions see extinct-
   ions/biodiversity loss
status anxiety, 131
steel, 40, 97, 195, 256, 258, 260, 261
stemming the flow, 185, 239-45
Stern, Nicholas, 100-1
stone, 49
Stone Age, 37, 50
Stuart, Tristram, 211
Styrofoam cups, 169
subsidies, agricultural, 211
substitutions, 193-4
sulphur compounds, 57
sulphuric acid, 57
Sumerians, 38
sun, 225, 226-7 see also solar
   energy
supermarkets, 206, 212-13, 290
supply push, 26-9
Sustain, 193
Sustainable Consumption
   Roundtable, 156
sustainable sources, 185, 187-99
Sweden, 52, 65, 119, 219, 257
Switzerland, 21, 64
synthetic fabrics, 76, 77, 190

tailings, 56
   dams, 57
tar sands, 111, 273
Tasmania, 64
taxation, 268, 284-6
technology, 37-41, 117, 213, 214,
   229, 230, 235-7, 246-52, 272-4,
   293

Tencel, 77
tennis rackets, 236
terbium, 113
textiles, 10, 33, 37, 45, 75-9, 82,
    190-1, 240-1
Thames, River, 101-2
Thames Survey Commission
    Board, 101
Thatcherism, 132
thermal mass, 195
thermal ployaspartate, 223
thin film technology, 250
3M corporation, 181
tidal power, 226
timber/timber trade, 61-3, 65-6,
    69, 154, 187, 195-6 see also
    paper; wood
tin, 38-9, 56
titanium, 113
toilet cisterns, 232
toilet paper, 12, 146-50
toilets
    composting, 219, 220
    flush, 216, 219
    see also toilet cisterns
toxic substances, 18, 71, 102-5,
    122
trade, terms of, 271
Trans-Amazonian Highway, 57
transformations, 37-41
tungsten, 180
TVs, 231-2
tyres, 79-80

UK/Britain
    Advisory Committee on
        Consumer Products and
        the Environment (ACCPE),
        155-6, 157
    and aggregates, 49, 54, 194
    and anaerobic digestion (AD),
        214, 215
    carbon emission targets, 100
    Climate Change Act, 123
    Commission on
        Environmental Markets

    and Economic Performance
        (CEMEP), 274-5
    Dark Ages, 39
    Department of Environment,
        Food and Rural Affairs
        (DEFRA), 151, 154-5, 156,
        283
    Department of Health, 99-100
    and disposable nappies, 221
    and energy, 228-9, 230, 232, 234
    and food waste, 211, 212, 214,
        215
    forests/woodland, 39-40, 61
    Gallagher Review on biofuels,
        209
    and glass, 52, 53-4
    and GM (genetic modifica-
        tion), 248
    government priorities, 275-6
    Industrial Revolution, 39-40
    industrial waste, 23, 82
    landfill, 18, 20
    and metals, 56, 58, 59-60, 112
    mining, 56, 58
    Ministry of Supply, 22
    new economics foundation
        (nef), 109, 115, 122
    and paper, 65-6, 67, 68
    peat, 150
    and phosphates, 143, 144, 145
    and plastic, 71, 72, 73
    and pollution, 101-2
    public procurement, 282-3
    and recycling, 22, 27, 53-4, 59-
        60, 63, 68, 72, 73, 82, 234,
        255-9, 260, 275-6
    and sanitation, 216, 217
    and second-hand textiles, 241
    Sustainable Consumption
        Roundtable, 156
    taxation, 284-5, 285-6
    and toilet paper, 147, 149
    Victorian age, 15
    vision of the future in, 8-12,
        266-72
    and water, 95, 96

and w...
wood-burni...
work of Gr...
    of bioplastics in, 206-7
    see also Eden Project; England;
        Scotland; Wales
UN Convention on Biological
    Diversity, 91
UN Environment Programme
    (UNEP), 61, 93
    Environmental Food Crisis
        report (2009), 94
uranium, 55
US, 65, 94, 101, 104-5, 113, 128,
    129, 147, 157-8, 158-9, 167-
    70, 211, 228
    National Academy of
        Sciences, 115
Uzbekistan, 98

VAT, 268, 285
vellum, 63-4
Venice, 52
Victorian age, 15, 129
Vienna Convention, 106
Vietnam, 62, 66
Viscose, 76, 77
Volta, Alessandro, 233, 234

Wackemagel, Mathis, 115
Wales, 265 see also UK/Britain
Walkers, 152, 279
'wants' and 'needs', 136-7
washi, 149
waste, 5, 6-7, 13-30
    dealing with see incineration;
        landfill; nutrients cycled;
        recycling; reuse
Waste and Resources Action
    Programme see WRAP
waste electonics and electrical
    equipment see WEEE
Waste Management Advisory
    Committee, 24
waste rocks, 56-7
water

...d agriculture, 77, 95, 96, 97-8
...ent civilisations, 38,
    ...
    and climate change, 95, 121
    and forests, 90
    good stewardship approach to,
        191-3
    and industry, 65, 82, 96-7
    and origin of life on earth, 35
    and pollution, 121
    and product standards, 278-9
    as resource, 89, 95-8, 118, 122
    and sanitation, 217, 220
water footprints, 192, 193
water-powered turbines, 228
wave machines, 228
Watson, J. D., 35
WEEE (waste electronic and
    electrical equipment), 260-
    2, 263, 296
    Directive, 6, 263
West Africa, 97
wetlands, 91
Which?, 155
wind power, 80, 226 see also wind
    turbines
wind turbines, 228, 229, 239
    blades, 80
Women's Institute, 277
wood, 39, 60-3, 77, 195-6, 225-6
wool, 75, 76, 77, 78, 190
Worboys, Nigel, 242
Worldwatch Institute, 149
WRAP (Waste and Resources
    Action Programme), 27, 53,
    122, 151, 157, 211, 215
WWF, 96, 209

Yellowstone, 169-70
yttrium, 55

Zambia, 241
zero waste, 171-5, 200
Zero Waste Academy, Kamikatsu,
    173
zinc, 104

www.vintage-books.co.uk